Mediterranean Perspectives

Series Editors
Brian Catlos
University of Colorado - Boulder
Boulder, CO, USA

Sharon Kinoshita
University of California Santa Cruz
Santa Cruz, CA, USA

As a region whose history of connectivity can be documented over at least two and a half millennia, the Mediterranean has in recent years become the focus of innovative scholarship in a number of disciplines. In shifting focus away from histories of the origins and developments of phenomena predefined by national or religious borders, Mediterranean Studies opens vistas onto histories of contact, circulation and exchange in all their complexity while encouraging the reconceptualization of inter- and intra-disciplinary scholarship, making it one of the most exciting and dynamic fields in the humanities. Mediterranean Perspectives interprets the Mediterranean in the widest sense: the sea and the lands around it, as well as the European, Asian and African hinterlands connected to it by networks of culture, trade, politics, and religion. This series publishes monographs and edited collections that explore these new fields, from the span of Late Antiquity through Early Modernity to the contemporary.

More information about this series at
http://www.palgrave.com/gp/series/15161

Gabriele Proglio
Camilla Hawthorne
Ida Danewid • P. Khalil Saucier
Giuseppe Grimaldi
Angelica Pesarini
Timothy Raeymaekers • Giulia Grechi
Vivian Gerrand
Editors

The Black Mediterranean

Bodies, Borders and Citizenship

palgrave
macmillan

Editors
Gabriele Proglio
Centre for Social Studies
University of Coimbra
Coimbra, Portugal

Camilla Hawthorne
Sociology and Critical Race and Ethnic Studies
University of California, Santa Cruz
Santa Cruz, CA, USA

Ida Danewid
Department of International Relations
University of Sussex
Brighton, UK

P. Khalil Saucier
Africana Studies Department
Bucknell University
Lewisburg, PA, USA

Giuseppe Grimaldi
University of Verona
Frontiera Sud Aps, Italy

Angelica Pesarini
New York University
Florence, Italy

Timothy Raeymaekers
Department of Geography
University of Zurich
Zürich, Switzerland

Vivian Gerrand
Alfred Deakin Institute for Citizenship and Globalisation, Deakin University
Melbourne, VIC, Australia

Giulia Grechi
Cultural Anthropology, Academy of Fine Arts Brera, Milan, Italy

Mediterranean Perspectives
ISBN 978-3-030-51390-0 ISBN 978-3-030-51391-7 (eBook)
https://doi.org/10.1007/978-3-030-51391-7

This Palgrave Macmillan imprint is published by the registered company Springer Nature Switzerland AG.
The registered company address is: Gewerbestrasse 11, 6330 Cham, Switzerland

Contents

Notes on Contributors

Ida Danewid is Lecturer in Gender and Global Political Economy at the University of Sussex.

Vivian Gerrand, PhD, is a research fellow at Deakin University's Alfred Deakin Institute for Citizenship and Globalisation.

Giulia Grechi, PhD, is Professor of Cultural Anthropology at the Academy of Fine Arts Brera, Milan, Italy.

Giuseppe Grimaldi, is a postdoctoral researcher at the department of social science of the University of Verona. He is the founder of Frontiera Sud Aps a project on the nexus between migration and territory in Southern Italy.

Camilla Hawthorne, PhD, is Assistant Professor of Sociology and Critical Race and Ethnic Studies at the University of California, Santa Cruz.

Cristina Lombardi-Diop, PhD, is Senior Lecturer in Modern Languages and Literatures and Women's Studies and Gender Studies at Loyola University, Chicago.

Angelica Pesarini, PhD, is a global lecturer at New York University, Florence.

Gabriele Proglio, PhD, is a research fellow at the Centre for Social Studies, University of Coimbra.

Timothy Raeymaekers, PhD, is a permanent research fellow at the Department of Geography, University of Zurich.

P. Khalil Saucier, PhD, is Professor of Africana Studies Department at Bucknell University.

List of Figures

CHAPTER 1

Preface

Cristina Lombardi-Diop

After 'the Mediterranean'[1]

What is the Mediterranean today: a solid sea or a liquid frontier, a bridge or a barrier? (Iain Chambers, "Maritime Criticism and Theoretical Shipwrecks")

Most reflections on the Mediterranean begin with an interrogation. In the unstable relationship between surface and depth, structure and process, unity and plurality, the Mediterranean Sea has often appeared as a paradox, an "endless epistemological provocation" (Lahoud 2013, 83), a site of both "postrepresentational understanding" and "critical depth" eluding definition (Chambers 2010, 679). Let's take the paradox of Fernard Braudel's monumental work, for instance.[2] Its methodology of the *longue durée*, while substantiating an argument in favor of 'the Mediterranean' as

[1] In this section, I will use 'the Mediterranean' in single quotation marks to signal an analytical category rather than the actual geographical sea.

[2] Fernand Braudel, *The Mediterranean and the Mediterranean World in the Age of Philip II*. Volume I and II.

C. Lombardi-Diop (✉)
Modern Languages and Literatures and Women's Studies and Gender Studies, Loyola University Chicago, Chicago, IL, USA
e-mail: clombardidiop@luc.edu

G. Proglio et al. (eds.), *The Black Mediterranean*, Mediterranean Perspectives, https://doi.org/10.1007/978-3-030-51391-7_1

a unitary sea, belies the fact that the actual unity of the Mediterranean world, if it ever existed, was disrupted in medieval times during the Crusades and again in the late sixteenth century, with the onset of the slave trade. It was then shattered by the consolidation of the European nation-state and its attendant imperialism in the nineteenth century. Writing at the time of the post-bellum reconstruction of Europe as a new order, Braudel acknowledges only marginally the pivotal role of the Mediterranean as a gateway to Europe's involvement with the transatlantic slave trade. In his complete *oeuvre,* comprising over 1300 pages, Braudel indexes the term "slaves" in one page, just to dismiss slavery in the West and East Mediterranean as "a curious attachment to the past." (Braudel 1995, 754) Yet, a few lines later, he writes that "slavery was a structural feature of the Mediterranean society" (Braudel 1995, 755) and not exclusive to the Atlantic and the New World. No other mentions of slavery follow after this statement. Concerned with the coherence of internal structures and the totality of natural history, his bird's eye-view of the Mediterranean basin traces the gradual fading of the line of palm tree, the vineyards, and the olive trees, mapping ecological discontinuities at the expense of human mobility. Braudel's Africa begins at the appearance of the sand.

The Black Mediterranean takes a different perspective as it plunges into the depth of the sea, concerned with its buried forces of racial terror and violence. In so doing, it dispels all illusions of structural ahistoricity. In making the Mediterranean Black, it begins to offer some responses to both the unitary vision and the postcolonial kaleidoscopic image of the sea as undecipherable, fragmented, fluid, and plural. As Michael Herzfeld reminds us, 'the Mediterranean,' as a unitary category, has often served belligerent cultural imperialism. It is enough to mention the Fascist and proto-fascist attempts at revitalizing the concept of *Mare Nostrum*. When translated into a heuristic device, it becomes a stereotype-making machine.[3] When contacts and conflicts are brought up to the surface (such as mercantile local economies, the clash between Christianity and Islam, piracy, and slavery in the post-classical period), the utopian vision of one

[3] Mediterranean stereotypical categories such as "honor" and "shame" still serve a series of self-representations in Southern Europe, as Herzfeld rightly asserts. I would add that categories such as "honor" and "shame," for instance, still support gender hierarchies in Italy, with male "honor" being one of the hidden structures of feeling behind the high numbers of femicide. See Herzfeld 2005.

Mediterranean fades into the recognition of many Mediterraneans, as David Abulafia suggests. Trade, for Abulafia, is what all Middle Seas share comparatively. It is trade that makes the Mediterranean paradigm expandable to other parts of the world (the Saharan Mediterranean Desert, the Mediterranean Atlantic, the Trans-Oceanic Caribbean, the Japanese Mediterranean, the Indian Ocean). Abulafia's structural model of 'the Mediterranean' interprets it as "an empty space between lands [...] in which waters create links between diverse economies, cultures, and religions" (Abulafia 2005, 65). His work opens new geographical perspectives, yet misses the opportunity to identify the common, extra-structural link (if trade, what type of global trade?) that connects these diverse systems of economic and social mobility. This is the fallacy that is inherent in the dominance of the classical, Mediterranean Studies model of the basin as a unitary sea. It is the de-linking of the Mediterranean from capitalist exploitation and its attendant infrastructures that used indentured racialized labor, restricted Black mobility, incarcerated racialized individuals in detention centers and camps, and denied them dwelling and citizenship rights as it happened in other systems of sea/ocean connectivity, and as it continues to do so in the contemporary Mediterranean. The proliferation of borders, from Tijuana to El Paso, from Malta to Lampedusa, from Bihaic to Idomeni, far from being 'exceptional,' makes manifest a mechanism of repetition of racial capitalist operations.

The Black Mediterranean works against the grain of previous formulations of this specific Middle Sea. By historicizing the Mediterranean as a *mare nero*, this volume puts the methodology of the *long durée* to good use, recognizing that contemporary forms of violence against Black migrants and refugees belong to the duration of racial subjection in Southern Europe. The historical role of Italy at the outset of the Atlantic slave trade is only one of the many connections between the Black Mediterranean and the Black Atlantic.[4] The "strong sequential links" (Davis 2000, 459) between the Atlantic slave system and the Italian Renaissance, involving Black slaves for the cultivation and the processing of sugar cane and the sugar production in late medieval Sicily and Andalusia, had their roots in Southern European medieval kingdoms, and

[4]A series of novels published in Italian by African writers chart the north-bound movement from Sub-Saharan and North Africa to Italy across the Mediterranean as a new Middle Passage. The apparition of ancestor spirits in these works indicate the presence of the afterlife of slavery and its temporal circularity. See Lombardi-Diop 2008.

in the slaveholding colonies in the eastern Mediterranean and on the coasts of the Black Sea (Blackburn 1997). When the people in the Mediterranean become visible as Black, the sea's historical trajectory appears punctuated and scarred, as Iain Chambers puts it, "by slavery, drowning, brutality, and the wrecked lives of ferocious migration today" (Chambers 2010, 681). The rise of European modernity and the ontological formation of humanity appears predicated on the narrative (and myths, fantasies, and imaginaries) of the inhumanity of Black subjects, what Achille Mbembe defines as "the Western consciousness of Blackness" (Mbembe 2017, 28).

The volume in its totality thus functions as a turning point in our critical thinking of migrations, borders, race, postcoloniality, and antiblackness. Its interdisciplinary, trans-historical approach to these areas of inquiry shifts the critical focus away from the contingencies of European national borders and the historical specificity of unilateral colonial/postcolonial relations. Asserting the need for "decolonizing the Mediterranean" (Proglio 2016) as a larger geopolitical and cultural area encompassing not only its norther and southern rims but also Sub-Saharan Africa, the volume inverts the epistemological trajectory of postcolonial inquiries. Instead of tracing the durable effects of European racial history in the Mediterranean present, it positions the contemporary Mediterranean "in the wake" (Sharpe 2016) of the afterlife of slavery as "a past that is not past" (Sharpe 2016, 13), and a practice whose modalities survive in the disregard for diasporic Black life in Southern Europe. It shows that the Mediterranean stages a 'crisis' that is not a state of exception, as biopolitical interpretations of the sea have repeatedly claimed, but a state of repetition of the subjection of Black life through the same old means: borderless apparatus of surveillance, containment, captivity, forced displacement, forced labor, the slave markets, and dehumanization. The separation between blackness and the heritage of European humanism precedes the material violence against Black bodies. As a corrective, the attendant responses to this predicament are also the same: fugitivity, wakefulness, citizenship struggles, rootedness, cultural transfigurations, and political subjectivity.

After the Black Atlantic

"[T]he Black Mediterranean is a variegated site of Black knowledge production, Black resistance and possibilities of new consciousness." SA Smythe ("The Black Mediterranean and the Politics of the Imagination").

In its analogy with the Black Atlantic, the Black Mediterranean as an analytical framework is less about migration and national paradigms and more about new transnational affiliations, the diffusion of transnational capital, and diasporic Black existence in the Mediterranean. As a diasporic framework the *Mediterraneo nero* allows the scholars gathered in this volume to re-center our critical attention on the forms of affirmation and resistance of Black life, addressing the response to the systemic and representational violence which affects Black people in Southern Europe. In the wake of the Black Atlantic, the volume explores the question of a double consciousness, also so central in Gilroy's work. However, in contrast to the cultural-studies turn inaugurated by *The Black Atlantic* (Gilroy 1993) the authors are less interested in literary, artistic, and cultural production and more in the configuration of everyday practices, life stories, fugitivity, antiblackness, and networks of solidarity mobilizing for civil rights. At least two of the chapters interrogate the predicament of Blacks as insiders and outsiders in Italian society, exploring the political and cultural implications of what it means to be Black Italian and Black European. Diasporic studies in a transatlantic context have shown scant interest in how the racial underpinnings of Italian migratory patterns have affected Blacks and non-white Italians. The word 'race' mostly refers to the differential treatment of Southern Italian migrants in the United States, yet often fails to connect such forms of unfree labor to the global outreach of racial capital.[5] While postcolonial scholars in Italy have explored—not exhaustively, but certainly productively—how race has defined whiteness, Mediterraneaness, and Italianness, the question is here reversed: what is the place of blackness within Italy and what happens when blackness and citizenship are considered mutually exclusive? What are we to make to of Mediterranean hybridity (*meticciato*) when it is evoked to neutralize blackness? This approach re-links the political struggles of Blacks in Europe to diasporic

[5] Donna Gabaccia's seminal work on Italy's global diasporas does mention the system of labor patronage in late 1890s New World run by the Italians as a form of unfree labor in a new global labor economy and places it along a continuum between slavery and free labor. See Gabaccia 2000.

spaces of struggle globally, such as the Baobab Experience in Italy, the #HungerForFreedom in the UK, the French *Parti des Indigènes de la Republique* (PIR), and the UK Black Lives Matter. Moreover, taking stock of the prolonged inertia of Leftist antiracist politics in their refusal to see African mobilization as a political struggle belonging to Italy, the volume specifically addresses the connivance between the system of both African and Italian *caporalati* (labor patronage) in the Southern fields and their imbrication with institutional and political constituencies at the expense of truly hybrid political spaces for Black militants in Southern Italy.

Why do we need this book today? Countering the reluctance to comprehend the Mediterranean 'humanitarian' crisis as a racial crisis beyond national borders, and to acknowledge the anti-racist struggles of refugees and migrants diasporically, ontologically, and trans-historically, the volume mobilizes the political potential of a new type of activist scholarship. Its authors represent a new generation of researchers, many of whom have first-hand experience or a family history of migration and dislocation, who are geographically dislocated across Europe and the United States, and who are trained in disciplines that are undisciplined and contaminated, such as cultural geography, cultural anthropology, international relations, cultural history, social history, or critical race theory. This is a truly collaborative effort of a collective of individuals, many of whom are activists speaking out in the public arena, demonstrating, mobilizing, questioning, and inaugurating a truly exciting moment of emancipatory scholarship in and around Italy.

References

Abulafia, David. 2005. Mediterraneans. In *Rethinking the Mediterranean*, ed. W.V. Harris, 64–93. Oxford: Oxford University Press.

Blackburn, Robin. 1997. The Old World Background to European Colonial Slavery. *The William and Mary Quarterly* 54 (1): 65–102.

Braudel, Fernand. 1995. *The Mediterranean and the Mediterranean World in the Age of Philip II*. Volume I and II. Trans. from the French by Siân Reynolds. Berkeley/Los Angeles: University of California Press.

Chambers, Iain. 2010. Maritime Criticism and Theoretical Shipwrecks. *PMLA* 125 (3): 678–684.

Davis, David Brion. 2000. Looking at Slavery from Broader Perspectives. *The American Historical Review* 105 (2): 452–466.

Gabaccia, Donna. 2000. *Italy's Many Diasporas*. Seattle: University of Washington Press.

Gilroy, Paul. 1993. *The Black Atlantic: Modernity and Double-Consciousness.* Harvard: Harvard University Press.

Herzfeld, Michael. 2005. Practical Mediterraneanism: Excuses for Everything, from Epistemology to Eating. In *Rethinking the Mediterranean*, ed. W.V. Harris, 45–63. Oxford: Oxford University Press.

Lahoud, Adrian. 2013. The Mediterranean, a New Imaginary. In *New Geographies, 5: The Mediterranean*, ed. A. Petrov, 81–100. Cambridge, MA: Harvard University Press.

Lombardi-Diop, Cristina. 2008. Ghosts of Memories, Spirits of Ancestors: Slavery, the Mediterranean, and the Atlantic. In *Recharting the Black Atlantic: Modern Cultures, Local Communities, Global Connections*, ed. Annalisa Oboe and Anna Scacchi, 162–180. London: Routledge.

Mbembe, Achille. 2017. *Critique of Black Reason.* Trans. and with an Introduction by Laurent Dubois. Durham/London: Duke University Press.

Proglio, Gabriele. 2016. *Decolonising the Mediterranean. European Colonial Heritages in North Africa and the Middle East.* Newcastle-upon-Tyne: Cambridge Scholars Publishing.

Sharpe, Christina. 2016. *In the Wake. On Blackness and Being.* Durham/London: Duke University Press.

Smythe, S.A. 2018. Suffering and the Limits of Relief. *Middle Eastern Report* 286 (Spring): 3–9.

CHAPTER 2

Introduction

Ida Danewid, Gabriele Proglio, Angelica Pesarini, Camilla Hawthorne, Timothy Raeymaekers, P. Khalil Saucier, Giulia Grechi, and Vivian Gerrand

The Mediterranean is once more confronting an important crisis of representation. Modern historiographies have often been characterized by essentialized interpretations, a kind of 'Mediterraneanism' (Herzfeld 2005) that continues to reproduce a necessary link between the Mediterranean's physical geographies and its, supposedly, distinct historical-cultural trajectory. While coming from different perspectives, these narratives share an implicit

I. Danewid
Department of International Relations, University of Sussex, Brighton, UK
e-mail: I.Danewid@sussex.ac.uk

G. Proglio
Centre for Social Studies, University of Coimbra, Coimbra, Portugal

A. Pesarini
New York University, Florence, Italy
e-mail: angelica.pesarini@nyu.edu

C. Hawthorne
Sociology and Critical Race and Ethnic Studies, University of California, Santa Cruz, Santa Cruz, CA, USA
e-mail: camilla@ucsc.edu

G. Proglio et al. (eds.), *The Black Mediterranean*, Mediterranean Perspectives, https://doi.org/10.1007/978-3-030-51391-7_2

cartographic vision of the Mediterranean space, one that reproduces its complex topography into a single, unitary epistemological framing and political management (Chambers 2008; Giaccara and Minca 2010). In our contemporary era, one central object of that epistemological frame is the Black African migrant. Seemingly floating on its fluid waters to reach the more stable topographies of continental Europe, Black migrants (categorized single-handedly as refugees, asylum seekers, displaced people) have come to increasingly symbolize the instability and insecurity of that space, which historiographers once described with such exoticizing and essentializing certainty.

Two metaphors appear to dominate this stereotype today: one which describes the African migrant essentially as a victim of stronger forces she or he cannot control; and another which describes the same subject basically as a criminal threat and a violent invader. Of the latter, an illustrative example has been the politics of closed borders initiated under during the so-called 'long Summer of migration' of 2015. After the rapid closure of Mare Nostrum, a humanitarian operation which saved thousands of lives, an increasingly repressive border regime in the Mediterranean pushed migrants towards taking other routes across the Balkans. Walls and fences quickly emerged on the Slovenian, Croatian and Austrian borders to push back migrants who were looking for protection. Across Europe, a rising xenophobic rhetoric accompanied the progressive militarization of borders and the reassertion of national territorial sovereignty. In Italy, the former Minister of Interior Marco Minniti as well as his successor Matteo Salvini have gained an impressive electoral consensus by exploiting the so called 'refugee crisis' with a focus on the 'emergency' of Black African migration. Like their counterparts in Austria, Hungary

---

T. Raeymaekers
Department of Geography, University of Zurich, Zürich, Switzerland
e-mail: timothy.raeymaekers@geo.uzh.ch

P. K. Saucier (✉)
Africana Studies Department, Bucknell University, Lewisburg, PA, USA
e-mail: pks008@bucknell.edu

G. Grechi (✉)
Cultural Anthropology, Academy of Fine Arts Brera, Milan, Italy

V. Gerrand (✉)
Alfred Deakin Institute for Citizenship and Globalisation, Deakin University, Melbourne, VIC, Australia
e-mail: vivian.gerrand@deakin.edu.au

and Poland, both have consciously instrumentalized the trope of African invasion to push through exceptional political measures, which dangerously infringe on basic human rights (amongst others by directly financing Libyan border guards who notoriously abuse migrant's fate) and which have dramatically reduced the possibilities of political asylum across Mediterranean waters. Akin to other sovereign 'strongmen' who have arisen in this epoch of crisis capitalism (Monbiot 2019), the main objective of such populist leaders is to exploit instability in order to distract the electorate from structural inequality and encourage the populace to blame migrants for their daily quandaries.

Of the former, a strong illustration has been the humanitarian response, which developed in response to the populist rise of neo-Fascism in contemporary Mediterranean migration politics. Invoking the primacy of the law of the seas as well as universal human rights to protect the victims of nationalist repression, a movement of civil disobedience has slowly taken shape these last years, which sails to the help of those migrants who continue to face life threatening conditions. A good illustration of this latter trend has been the public reaction to the disobedience of Carola Rackete, a German ship captain who picked up 53 migrant people in the Mediterranean at a few miles distance from the Libyan shore, to bring them to safety. Rejecting the offer by the Libyan authorities to dock at Tripoli because of the horrible human rights conditions in the country, her Sea-Watch 3 steadily moved toward Lampedusa in the days of June 2019. After the refusal by the Interior Minister Salvini to open Italian ports, she decided to force the blockade. She was arrested by police, accused, and later acquitted, of facilitating illegal migration.

While standing diametrically opposed in ideological terms, both stereotypes share a similar, whitening gaze. By reproducing basically voiceless subjects that are both talked about and talked for, this gaze simultaneously reproduces the conditions of their very exclusion from the dominant framings of human rights, political asylum and territorial citizenship discourse. To some readers, this parallel may sound like an anomaly. To the authors of this volume it does not. Rather, we argue that the contemporary framing of the Mediterranean crisis subscribes itself to a historicizing, white, and predominantly male European gaze, which continues to frame its excluded others as either 'charitable subjects' or 'uninvited guests' whose histories and trajectories are consequently erased (Saucier and Woods 2014; Danewid 2017). For us, and many others beyond this volume, what is unfolding in the Mediterranean today is, in part, constitutive of an unparalleled crisis—not a crisis of humanitarianism or of territorial sovereignty, but of Eurocentrism. While migration indeed remains a complex and dynamic phenomenon where routes, countries of origin and transit, trajectories, and mobilities constantly change, the Blackness of the Mediterranean as a topological intersection of entanglements, trajectories, and hierarchies is becoming ever more visible through the silenced voices of those who dare speak out for themselves. Anti-Blackness serves not only to confront these emerging subjectivities; but also to continue framing the rights of citizenship and belonging into a dominant, white European legislative and ideological frame.

Moving explicitly against such objectifying, white perspective of the distant—be it caring or protecting—observer, the authors of this volume share an urge to give space to an alternative narrative, which simultaneously dislocates this whitening gaze, and which, we hope, may offer scope for a Black politics of reclaiming Mediterranean space.

In today's humanitarianist and territorialist frames, Blackness, more often than not, continues to function as the symbol, the metaphor, and the trope of either criminality and violence, or of victimhood and harmless pity—similarly to the colonial 'savage' in European political thought. In the context of contemporary Mediterranean migration, these tropes have been increasingly present in the transformation of today's border politics. Not surprisingly, the 'long Summer of migration' has been likened many times to a kind of Sophoclean drama, in which the laws of humanity are pitted against those of sovereign, despotic rule (Raeymaekers 2014). Depending on the ideological spectrum, nationalist despotism or humanitarian protection take turns in replacing the voices and presences of those who are cared for or protected against. This typology of story tale positions the subject internally to the hegemonic discourse of classicism and has several echoes in the postcolonial condition, in particular, in the positionality of those in the act of positioning those subjects: white, woman, man, European, middle-class, acting to help migrant people or protect threatened citizens against them. This ontology of 'spoken for' subjects reproduces a strong, forensic epistemology that simultaneously displaces the point of view of the living, and replaces the memorization of the death.[1]

[1] Another, illustrative example of this forensic gaze has been the artistic recovery of the shipwreck by Swiss artist Christoph Büchel. The wreck, which sank on April 18, 2018 in front of the island of Lampedusa, contained the bodies of 452 dead bodies; a total of over 900 people who died on that boat; only 28 migrants survived. Christoph Büchel, who is known for his provocative installations on migration issues, hoisted the rusty wreck onto a lorry trailer just in time for the opening of the Art Biennale in Venice on May 11, 2019, where it has the ambition to serve as a memorial of migrant deaths. In the meantime, a forensic team of scholars has continued to dig up human remains. After several years of research, only two young Mauretanians have been identified.

To summarize, we argue in this book, the Mediterranean continues to show different shades of white, if white is the color of the hegemony and normativity. Within the dominant epistemology of a Eurocentric gaze, this white Mediterranean recalls the theoretical framework proposed by Gayatri Chakravorty Spivak, who wrote about the case of *sati* and the Indian woman who cannot speak because of her being silenced both by the patriarchy and colonialism (Spivak 1988). According to Spivak's reflection, those subjects are subalterns and cannot speak or, in some cases, are ventriloquized—or spoken for. The theme of representation—or rather representativeness—shows powers and their typologies of actions in the production of the social space. In this sense, the Mediterranean can be interpreted as a complex, dense and deep network of intersected and intertwined narratives. To untangle this skein requires carefulness and precision in the deconstruction process, paying attention to each archive mobilized in producing a discursivity (image, imaginary, visual representation) and to its use for reaching specific goals according to positionalities and aims of the speakers' subjects. In addition to this, it is important to take into account that the scholar is always "working backlight", namely the representation is the output of a narrative process based on knowledge and power that makes visible a certain view of the world. Who is represented is de-subjunctivized and appears with a mask: the subaltern cannot speak because she or he is incapsulated in a word—its meaning and connected practices of dominance—image or view of the world. Deconstructing a representation cannot be separated from a political interpretation about the act of naming. In this sense, "working backlight" means to be aware that beyond and behind the representation there can be forms of resistance by the same subject who is represented. According to this statement, the Black Mediterranean is not only a physical place which excesses the geographical space of the sea (Proglio 2018) and concerns the act of being in diaspora; it is a practice of decentralization of the gaze and the production of knowledge from representation to resistance in opposition to a canonized and uniformized geography of power; it is a practice of investigation through which the chronological continuity of history is interrupted by unwritten and unsaid personal and collective stories about struggles against colonial, national, European, patriarchal and white powers; it is a practice of scrutinizing archives beyond their organization and modes of operation, starting from the historical source in order to pay attention to supposed silence or what is yet unsaid.

The Black Mediterranean

Inspired by Robert Farris Thompson (1984/2010) and Paul Gilroy's (1993) theorizations of the Black Atlantic, the Black Mediterranean captures the long history of racial subordination and resistance in the Mediterranean region. In some of the earliest texts addressing the Black Mediterranean, scholars such as Cedric Robinson (1983/2005) and Robin D. G. Kelley (2005) understood the Mediterranean as both a precondition for modern racial capitalism *and* as a site for the ongoing reproduction of regimes of racialization and Black subjectivities. Historians have long recognized the Mediterranean Sea as a unique space, one that has brought three continents into direct contact and facilitated dense networks of cross-cultural exchange, commerce, and mobility since the times of antiquity (Abulafia 2011; Diop 1989; Braudel 1949/1995; Broodbank 2013; Chambers 2008; Horden and Purcell 2000). But while they acknowledge this deep history of transculturation (Pratt 1992/2008), scholars of the Black Mediterranean also challenge the romanticization of the Mediterranean as a space of convivial exchange and unfettered hybridity, pointing to oft-overlooked histories of racial violence and their contemporary reverberations (Giaccara and Minca 2010; Harris 2005; Lahoud 2013).

As an analytic, the Black Mediterranean approaches the Mediterranean Sea as a space of multiple mobilities, traversed by various frontiers and border technologies, and spanning both colonial legacies and postcolonial conditions. In this edited volume, we employ the Black Mediterranean as a framework for addressing the role of the "colour line" (Du Bois 1903/2007) in defining forms of exclusion and differential inclusion across the many shores of the Mediterranean. In other words, Blackness is connected to the symbolic and discursive instruments of a captive society, in addition to the physical instruments used to control Black people. It engenders greater impunity under the cover of law and the legislative framework for policing the color line. More specifically, we explore, at times, the ways in which anti-Blackness inheres in the production and creation of legislation and jurisprudence itself, despite legal progress in certain domains of life. Perfected post-racial forms of surveillance have become a required juridico necessity and legal creations help structure the modern dynamics of legality and guideposts that help maintain and fortify the coherence and material continuation of blackness as lack, as negation: that is Blackness as criminal. In this sense police power is both structural and cultural. This intervention builds upon and responds to the work of critical border and migration studies in Europe (see, for instance, De Genova 2017; Jones 2016; Mezzadra and Neilson 2013)—a scholarship that has generated timely and politically urgent insights about the relationship between exclusion, capitalism, and differential regimes of mobility across the Mediterranean region. Yet much of this work has tended to elide a more systematic analysis of racism and anti-Blackness as *integral* to the dynamics of trans-Mediterranean migration, exclusion, and differential incorporation.

As such, this volume seeks to rethink the contemporary European migrant crisis in the Central Mediterranean by deliberately foregrounding questions of race and Blackness. This approach is necessitated by the sheer material realities of Black life and death across the Mediterranean. A significant proportion of the migrants, refugees, and asylum-seekers who are attempting to enter Europe via the central Mediterranean route are traveling from home countries in sub-Saharan Africa.[2] In addition, Black migrants endure disproportionate levels of violence, abuse, and danger at every stage of their journeys—including upon their arrival to Europe. And finally, the figure of the Black refugee (more so than any other migrant group) is routinely discursively mobilized by the European far-right as representative of the "threats" of invasion, racial contamination, and ethnic substitution.

As noted above, Paul Gilroy's *Black Atlantic* [1993] serves as both a theoretical reference point for this project, and also represents an injunction to identify the specific conditions of the Black diaspora in the Mediterranean—a space that has been shaped by deep histories of colonialism, enslavement, and migrations between Africa to Europe. Gilroy argued that the Black Atlantic constituted a 'counterculture of modernity' that was grounded in the foundational rupture of the trans-Atlantic slave trade, thus throwing into question the racial pretensions Euro-American ethnic absolutism. The Black Mediterranean can similarly generate a radical re-telling of "Europeanness" by locating questions of race and Blackness at the center of the region typically understood to be the "cradle" of European civilization. The Black Mediterranean points to the longue durée of Blackness in Europe which, as Stephen Small (2018) argues, constitutes the physical presence of people of African descent in Europe,

[2] See https://data2.unhcr.org/en/situations/mediterranean/location/5205 for more information.

forms of "race-thinking and racist thinking," and the intangible Black cultural presence (often used by Europeans to reinforce stereotypes of African savagery and Black inferiority).

At the same time, we are careful not to equate the Black Mediterranean and the Black Atlantic, or to uncritically superimpose Black Atlantic histories and frameworks of racialization onto the Mediterranean. After all, the contemporary Mediterranean refugee crisis is not identical to the Middle Passage, there are distinct histories and legacies of colonialism in the Mediterranean, and the politics of Blackness in Europe are often articulated differently than in the Americas (see Hine et al. 2009). However, we do acknowledge that racial slavery has left an indelible mark on the history and structure of (non) recognition of those officially and legally invited to the fraternal fellowship of personhood and those indefinitely uninvited: that is the legal archive of racial slavery casts light, albeit not total light, on the distinction between blackness and personhood globally.[3]

As Stuart Hall argued, there is not one racism in general but multiple *racisms* (Hall 1986), and the Black diaspora is characterized not so much by uniformity as by heterogeneity and diversity (Hall 1990). The Black Mediterranean thus provides the opportunity to provincialize North American approaches to the study of Blackness (Wright 2015). But while we situate our analyses in historical and geographical specificity, we also acknowledge that the Mediterranean is not exceptional—rather, it remains connected to the global and world-historical scales of anti-Black racism. After all, it is the collective violence of racism (a "death-dealing displacement of difference into hierarchies that organize relations within and between the planet's sovereign political territories" [Gilmore 2002]) that produces the ever-shifting, objective reality of "race" (see Viswewaran 2010) across different historical and geographical contexts.

We also employ the Black Mediterranean in this volume as a way to emphasize Black resistance. Instead of talking about subalternity—and the role of power in controlling bodies and mobilities across the Mediterranean—we are interested in documenting forms of resistance. The Black Mediterranean, in this sense, emerges as a sea with an uncertain and ongoing changeable geography, which does not correspond with that

[3] See for example Charles Mills, *The Racial Contract*. Cornell University Press, 2014; Sylvia Wynter, S. Unsettling the Coloniality of Being/Power/Truth/Freedom: Towards the Human, After Man, Its Overrepresentation–An Argument. *CR: The New Centennial Review* 3, 3 (2003): 257–337. See also DuBois.

produced by nation-state divisions and forms of spatial organization from the supposed centrality of Europe. In fact, to lose the reference points of a compass that is always oriented to the North means to observe how resistances to the hegemony are strictly connected with a knowledge production, which happens without respect for the order of discourse (to borrow a formula coined by Michel Foucault [1971]).

The Black Mediterranean, then, reveals a transnational geography based on those Black people who experienced, directly or indirectly, various typologies of violence based on the process of naming—through the colonial archive (Stoler 2010; Wekker 2016), with its taxonomy, categories, representations, and practices. Forms of resistance and resilience, strategic practices of disguise and alteration of Self in order to reach a goal—to cross the border or to elude citizenship devices that produce both for exclusion and differential inclusion—are, all of them, part of the Black Mediterranean. In this sense, it is possible to talk about the black of the sea in terms of what Edouard Glissant theorized as the "right to opacity" (2009), or as bodies out of place (Puwar 2004), with subjectivities able to exceed the norm, transgress the role of dichotomies, and invent ways to take advantage of imposed positionalities.

Finally, the Black Mediterranean represents a powerful ethical-political demand. With regard to the role of memory, the Black Mediterranean works as an archive in order to collect, share, and re-actualize the teaching of positive and negative attempts to violate, break, and disobey the norm—both the law as well as social, ethical and moral conventions. At the same time, this archive of memory harks back past stories—such as those concerning colonialisms and slavery—in order to draw common roots for subjects who have suffered a geographical, cultural, and emotional displacement. From an historical point of view, then, the Black Mediterranean evokes a conceptualization of time that is not only linear, circular, and continuous: in fact, it makes evident how the production of historical knowledge is based on a canonized idea of time and temporality. As pointed out by Iain Chambers, it is possible to discover discontinuities and interruptions, a bending progression of the chronological trend, and a sectoral categorization of what is and has been known as past, present and future (Chambers 2017, Chambers and Cariello 2019).

Theoretical and Methodological Scope

One ambition of this volume is to make more explicit what Merrill (2013) and McIntyre and Nast and (2011) call the racial geography of Black subordination across Afro-European borders. Taken together, the Black Mediterranean highlights the current re-configurations of postcolonial capitalism under conditions of late liberalism, which, as Marion Werner indicates, "indexes the ways that capitalist accumulation is constituted through the reworking of hierarchies of racialized and gendered difference, thus redrawing the social and spatial boundaries between hyper-exploited wage work and the people and places cast out from its relations" (Werner 2011; see also Quijano 2000; Mezzadra and Neilson 2013).

On the one hand, our concern lies with the constant reworking of such differences through multiple operations of anti-Blackness and racial violence, of gendered and racial capitalism. For example, in the Italian context, different chapters refer to the ways in which subsequent governments of Matteo Renzi and Matteo Salvini have consistently contributed both legally and mediatically to the transformation of immigrants into illegalized subjects, as *clandestini*, invaders and rapists who are reproduced both as a formidable resource for the country's underground economy and as a profitable scapegoat for its endemic political crisis. While not exclusively targeting Black people, the imagery of desperate African refugees who cross the Mediterranean waters in despair serves to bolster this mechanism of simultaneous exclusion and enslavement. In their assessment of the Mediterranean middle space, both Renzi and Salvini have consistently made it clear that the responsibility for today's emergency lies with Libyan and other foreign 'slave traders', NGO's and traffickers, and not with European capitalist enterprises whose reproduction depends on cheap racialized labor. Paradoxically, therefore, the active association of clandestine migrant subjects with slavery, exclusion and humanitarian emergency serves to reproduce such racial formations as a structural element of European capital accumulation.

On the other hand, our reflections resonate with Frantz Fanon's observations of being "black in relation to the white man" (1952/2008) in Europe. In his experience, he writes about encountering the consciousness of his own black body as a negation, a "third person consciousness." The various contributions of this volume narrate how the presence of Black people in postcolonial capitalism are actively woven, through "a thousand details, anecdotes, stories", into a white hegemonic narrative about what Blackness means in relation to white capital and production, and how this hegemony consistently throws blackened subjects into a feeling of imprisonment and dislocation, by taking people far off from their own presence.

The Black Mediterranean Collective

We believe that an inherently interdisciplinary analytic such as the Black Mediterranean necessarily requires a multi-vocal, collaborative, and non-hierarchical approach to editing. This format is inspired by a long history of collective editorial work in academia, including the path breaking volumes published by the Centre for Contemporary Cultural Studies (including *The Empire Strikes Back: Race and Racism in 70s Britain*). Collective writing creates a greater capacity for integrative production. This collective enterprise is not so much because of an engagement with the problem(s) at hand, more so what has emerged from the problem. We, like Cedric Robinson's (1983/2005, xxx) cogent observation, are an "accretion… of collective intelligence gathered from struggle." The problem at hand is so massive and intense that to do it alone is inadequate. Writing in common creates the condition of possibility for forming a people that does not act merely as a group of individuals, scholars that know, but as a collective. Writing together and with each other is understood as a praxis of liberation. It is our way of gathering in order to contest and insist on political imaginaries beyond the white, Eurocentric West.

Further, writing together, as a collective, highlights the differences in style, technique, and criticism, without reproducing hierarchies of intellectual power. For while we all see a common problem, we all have a different approach that speaks to a diversity of thinking and acting. Aimé Césaire (2010, 152) best sums up the purpose of the collective when he writes that it is "enriched by all that is particular"; indeed, a "enriched by every particular, the deepening and coexistence of all particulars," not to mention that the collective involves scholars based in different universities around the world, with different research paths and methodological approaches but with the same aim to analyse, deconstruct and disassemble those bodies from which the Eurocentrism is thought, imagined and made real.

Overview of the Volume

The chapters in this book are organized according to three overarching themes: borders, bodies, and citizenship. Bringing together scholars working in geography, political theory, sociology, and cultural studies, this volume takes the Black Mediterranean as a starting point for addressing a set of crucial questions about the racialized production of territories, borders, bodies, subjectivities, and citizenships in contemporary Europe: What is the role of borders in controlling migrant flows from North Africa and the Middle East? What is the place of Black subjects in the Central Mediterranean migratory context? And what is the relevance of citizenship (in its many forms) to discussions about the possibilities of Blackness and Black life in Europe?

Borders are both material and symbolic; they enclose the geopolitical boundaries of nation-states, but they also extend far beyond it and reach inside it, pervading even the most mundane experiences of everyday life. In addition, numerous scholars have demonstrated that borders cannot be understood separately from histories of colonialism and slavery—modern technologies of bordering and surveillance have their roots in the violent practices emerged to manage the movement of colonized and enslaved populations (see Browne 2015; Hom 2019). These sorts of critical historical accounts are of crucial political importance—by challenging the presentism that characterizes mainstream analyses of migration "crises," they in turn force us to question taken-for-granted notions of borders as fixed, natural, and stable. In "When the Mediterranean became Black," Angelica Pesarini takes on this task by charting a genealogy of the Mediterranean as a political category, and its relationship to the ongoing construction of Italianness. Starting with Matteo Salvini's discursive externalization of the Mediterranean as a source of invasion and racial contamination, her chapter then turns to a consideration of the various ways "Mediterraneanness" was deployed during Italian national unification, colonialism, and fascism to produce and contest the racial boundaries of Italianness and citizenship.

While the category of the Mediterranean has been deployed to calcify exclusionary notions of Italianness, a growing number of activists and artists have also attempted to use the Mediterranean *transgressively*, in ways that challenge national borders and the violent colonial histories through which they were produced. In "Colonial Cultural Heritage and Embodied Representations," Giulia Grechi considers the ways that contemporary artists in Italy such as Martina Melilli and Tania El Khouri are working to deconstruct racist colonial imaginaries in the Mediterranean and, in doing so, denaturalize the borders of Italy and Europe. These artists, she argues, directly confront the ongoing production of ignorance (Mills 2007) about Italian colonialism—what she calls the "colonial unconscious"—by drawing connections between colonial systems of representation and the current spectacularization and Othering of Black migrants crossing the Mediterranean today.

Gabriele Proglio's chapter similarly looks to the Black Mediterranean's colonial archive. In "Fanon in the Black Mediterranean," Proglio engages with anti-colonial intellectual Frantz Fanon to analyze the condition of "the wretched of the Mediterranean"—the thousands of Black migrants positioned at the borders of Europe. Drawing on Fanon's theorization of violence, he considers the many symmetries that connect the conditions of the colonized "natives" in Fanon's work to the lived experiences of migrants crossing the Mediterranean today. Drawing on political analysis, media representation, and the voices of migrants themselves, Proglio demonstrates that the European border regime has produced a Manichean world, one that bears a direct historical relationship to the racist colonial landscapes condemned by Fanon.

The next set of essays in this volume addresses the theme of bodies. But rather than uncritically capitulate to the description of Black migrants as fungible bodies,[4] as they are so often reduced in media representations and political rhetoric, these chapters consider the range of power-laden process by which Black *people* are reduced to Black *bodies*—or, in the words of Hortense Spillers (1987), mere *flesh*. In what ways do processes of border violence, segregation, and racial-capitalist exploitation depend upon the production of expendable, excess racialized bodies? And what possibilities remain for the emergence of "a true humanism—a humanism made to measure the world" (Césaire 1950), one that that confronts the racist exclusions inherent to the category Sylvia Wynter called "Man" (Wynter 2003)? P. Khalil Saucier's chapter in this volume considers the ways that the ontologization of Blackness as outside the category of "Man" or "human" complicates liberal humanitarian interventions in the

[4] To quote Katherine McKittrick (2016, 4), "[W]e must ask ourselves…how black *bodies* rather than *black people* are informing how we understand the production of space and the production of knowledge and, as well, how these bodies that tidily uphold our academic ideas inadvertently or explicitly replicate a biocentric order."

Mediterranean. “Carne Nera” begins by describing a grainy undercover video that began to circulate in the fall of 2017 of a modern-day Libyan slave market. The video ignited an international firestorm, prompting “new abolitionists” to respond with a range of progressive, antiracist gestures and actions. But despite being stimulated by black death in the Mediterranean and the auction blocks of North Africa, Saucier argues that this neo-abolitionism says very little about racism and antiblackness. He contends that neo-abolitionism in the Mediterranean remains invested in the category of the “human” as an all-encompassing (and non-racial category) for thought and praxis, rather than, as Fanon forcefully posited, “the wretched of the earth” who compel us to rethink traditional discourse and taken-for-granted categories.

Timothy Raeymaekers’ contribution turns from questions of ontology and ontologization to an analysis of the relationship between bodies, forms of spatial organization, and political subjectivity. In “Impermanent Territories,” he engages with critical urban and cultural studies to explore the emergence of a “black African ghetto” in the midst of reconfiguring Mediterranean geopolitics. Raeymaker’s chapter focuses specifically on southern Italy, where the intensification of refugee flows and the precarization of migrant lives during the past five years has led to a growth of migrant ghettos—spaces that, he argues, are built on a particular construction of race, of urban space, and sovereign exceptionalism. They serve both as waystations for migrants en route to destinations further afield, and as more permanent clusters for migrant workers in the agricultural sector. These ghettos are characterized by a “persistent ambivalence”: on the one hand, they are the product of spatial segregation and marginalization, and on the other hand, they have also become sites of struggle from which Black migrants actively constitute new political subjectivities.

In “These Walls Must Fall,” Ida Danewid considers another way in which racialized bodies are governed through modes of spatial configuration—specifically, the range of colonial, racial capitalist strategies for regulating mobility. Starting from the seeming paradox that globalization has gone hand in hand with an intensification of border fortification, Danewid argues that European capitalist modernity has relied upon the racialized control of mobility since its inception. Enslaved Africans, indentured servants, sharecroppers, and contemporary “economic migrants” represent different historical and geographical iterations of the “highly expendable, super-exploitable, and *moveable* laboring subjects” necessary for capital accumulation. Bringing the Black Mediterranean into conversation with the Black Radical Tradition's emphasis on the relationship between racism and capitalism, Danewid challenges the political production of distinctions between "immigrants" and the "white working class" and suggests that any antiracist and anticapitalist political program must also call for the eradication of borders.

The third set of essays in this volume are concerned with citizenship. While discussions of the Mediterranean migration 'crisis' have been primarily concerned with the act of *crossing*, less attention has been paid to processes of settlement and territorialization. Yet racial violences continue even after the geopolitical border has been crossed—for instance, through restrictive citizenship laws that affect the children of immigrants who themselves never actually migrated across a physical, geopolitical border. By turning to an analysis of citizenship, this final set of essays considers the multiplication of geographies, subjectivities, and temporalities across the Black Mediterranean. In "L'Italia Meticcia?" Camilla Hawthorne uses the Black Mediterranean as a framework to explore unfolding debates over the boundaries of Italian citizenship. She argues that the deep entanglements of race and citizenship signaled in ongoing debates about the so-called *seconda generazione* [the Italian-born children of immigrants], *jus soli* [right of birthplace citizenship], and Black Italianness are inflected by a long history of anxiety and contestation over the racial status of Italy, perched as it is at the geographical and metaphorical edges of Europe and Africa. Hawthorne argues that questions of who truly "belongs" in Italy as a rightful citizen continue to be shaped by the much-hyperbolized threat of African contamination and Italian racial degeneration.

In "Re-Imagining Citizenship in the Black Mediterranean," Vivian Gerrand also investigates the ways that dominant conceptualizations of Italian citizenship are reproduced through racially exclusionary narratives, images, and legal structures. Her chapter examines the ways that citizenship claims from Black Italians—particularly those with heritage from Italy's former African colonies—radically challenge and transform notions of Italian national belonging. Gerrand argues that novel Black Italian literary, political, and historical engagements with citizenship—especially those that stress the mutual constitution of Italy and its former colonies—can challenge bounded notions of Italian identity, belonging, and membership.

As the contributors to this volume argue, a critical engagement with Italian colonial legacies provides insights that can be used to challenge enclosed, ethnic absolutist (Gilroy 1993) understandings of Italianness as they are enacted in violent practices of border fortification and exclusionary citizenship laws. In "The *Habesha* Italians," Giuseppe Grimaldi focuses on practices of identification among the "second generation" members of Eritrean and Ethiopian diaspora in Italy. His chapter, based on ethnographic research conducted in Milan's Porta Venezia neighborhood, considers the ways in which Italians of Eritrean or Ethiopian origins work with and against hegemonic understandings of Italianness, activating powerful new political imaginaries and practices. In particular, Grimaldi explores the deployment of "*Habesha*-ness" as a social category that indexes Eritreans and Ethiopians—former Italian colonial subjects—as simultaneously inside and outside the boundaries of Italianness, as well as signaling their connection to a global Black diaspora.

In the chapters of this volume, the Black Mediterranean forms a counter-archive that encompasses memories of resistance during colonialism and in the postcolonial period; historically- and geographically-distinct processes of racial formation and articulations of racism; overlapping diasporas and migratory journeys; and practices and lived experiences of Blackness in Europe. The Black Mediterranean, we emphasize throughout, is not only a catalog of Black death and abjection; it is also a site that conserves transgressive practices of Black space-making, creative forms of resistance, and new subjectivities.

References

Abulafia, David. 2011. *The Great Sea*. Oxford University Press.

Braudel, Fernand. 1949/1995. *The Mediterranean and the Mediterranean World in the Age of Philip II*. Reprint. Berkeley: University of California Press.

Broodbank, Cyprian. 2013. *The Making of the Middle Sea: A History of the Mediterranean from the Beginning to the Emergency of the Classical World*. Oxford: Oxford University Press.

Browne, Simone. 2015. *Dark Matters: On the Surveillance of Blackness*. Durham: Duke University Press.

Césaire, Aimé. 1950/2000. *Discourse on Colonialism*. Reprint. New York: Monthly Review Press.

———. 2010. Letter to Maurice Thorez. *Social Text* 28 (2): 145–152.

Chambers, Iain. 2008. *Mediterranean Crossings: The Politics of an Interrupted Modernity*. Durham: Duke University Press.

———. 2017. *Postcolonial Interruptions, Unauthorised Modernities*. London: Rowman & Littlefield.

Chambers, Iain, and Marta Cariello. 2019. *La questione mediterranea*. Milan: Mondadori.

Danewid, Ida. 2017. White Innocence in the Black Mediterranean: Hospitality and the Erasure of History. *Third World Quarterly* 38 (7): 1674–1689. https://doi.org/10.1080/01436597.2017.1331123.

De Genova, Nicholas. 2017. *The Borders of "Europe": Autonomy of Migration, Tactics of Bordering*. Durham: Duke University Press.

Diop, Cheikh Anta. 1989. *The African Origin of Civilization: Myth or Reality*. Chicago Review Press. ISBN 1613747365.

Du Bois, W.E.B. 1903/2007. *The Souls of Black Folk*. Reprint. Oxford: Oxford University Press.

Fanon, Frantz. 1952/2008. *Black Skin, White Masks*. Reprint. New York: Grove Press.

Foucault, Michel. 1971. *L'Orde du discourse*. Paris: Gallimard.

Giaccara, P., and C. Minca. 2010. The Mediterranean Alternative. *Progress in Human Geography* 35 (3): 345–365.

Gilmore, Ruth Wilson. 2002. Fatal Couplings of Power and Difference: Notes on Racism and Geography. *The Professional Geographer* 54 (1): 15–24.

Gilroy, Paul. 1993. *The Black Atlantic: Modernity and Double Consciousness*. Cambridge, MA: Harvard University Press.

Glissant, Édouard. 2009. *Philosophie de la relation. Poésie en étendue Phases XI. La pensée de l'opacité du monde*. Paris: Editions Gallimard.

Hall, Stuart. 1986. Gramsci's Relevance for the Study of Race and Ethnicity. *Journal of Communication Inquiry* 10 (2): 5–27. https://doi.org/10.1177/019685998601000202.

———. 1990. Cultural Identity and Diaspora. In *Identity: Community, Culture, Difference*, ed. Jonathan Rutherford, 222–237. London: Lawrence & Wishart.

Harris, W.V., ed. 2005. *Rethinking the Mediterranean*. Oxford: Oxford University Press.

Herzfeld, M. 2005. Practical Mediterraneanism: Excuses for Everything, from Epistemology to Eating. In *Rethinking the Mediterranean*, ed. W. Harris, 45–63. Oxford: OUP.

Hine, Darlene Clarke, Trica Danielle Keaton, and Stephen Small, eds. 2009. *Black Europe and the African Diaspora*. Urbana: University of Illinois Press.

Hom, Stephanie Malia. 2019. *Empire's Mobius Strip: Historical Echoes in Italy's Crisis of Migration and Detention*. Ithaca: Cornell University Press.

Horden, Peregrine, and Nicholas Purcell. 2000. *The Corrupting Sea: A Study of Mediterranean History*. Malden: Blackwell Publishing.

Jones, Reece. 2016. *Violent Borders: Refugees and the Right to Move*. London: Verso.

Kelley, Robin D.G. 2005. Foreword. In *Black Marxism: The Making of the Black Radical Tradition*, by Cedric J. Robinson (1983; reprint, Chapel Hill: University of North Carolina Press), xi–xxvi.

Lahoud, Adrian J. 2013. The Mediterranean, a New Imaginary. In *New Geographies, 5: The Mediterranean*, ed. A. Antonio Petrov, 81–100. Cambridge, MA: Harvard University Press.

McIntyre, M., and H.J. Nast. 2011. Bio (necro) Polis: Marx, Surplus Populations, and the Spatial Dialectic of Reproduction and "Race". *Antipode* 43 (5): 1465–1488.

McKittrick, Katherine. 2016. Diachronic Loops/Deadweight Tonnage/Bad Made Measure. *Cultural Geographies* 23 (1): 3–18.

Merrill, Heather. 2013. Who Gets to Be Italian? Black Life Worlds and Black Spatial Imaginaries in Turin. In *Geographies of Privilege*, ed. France Winddance Twine and Bradley Gardener, 135–160. New York: Routledge.

Mezzadra, Sandro, and Brett Neilson. 2013. *Border as Method, or, the Multiplication of Labor*. Durham/London: Duke University Press.

Mills, Charles. 2007. White Ignorance. In *Race and Epistemologies of Ignorance*, ed. Shannon Sullivan and Nancy Tuana, 11–38. Albany: SUNY Press.

Monbiot, George. 2019. From Trump to Jonhson, Nationalists Are on the Rise—Backed by Billionaire Oligarchs. *The Guardian*, July 26. https://www.theguardian.com/commentisfree/2019/jul/26/trump-johnson-nationalists-billionaire-oligarchs

Pratt, Mary Louise. 1992/2008. *Imperial Eyes: Travel Writing and Transculturation*. 2nd ed. Reprint. New York: Routledge.

Proglio, Gabriele. 2018. Is the Mediterranean a White Italian-European Sea? The Multiplication of Borders in the Production of Historical Subjectivity, in Interventions. *International Journal of Postcolonial Studies* 20 (3): 406–427.

Puwar, Nirmal. 2004. *Space Invaders: Race, Gender and Bodies Out of Place*. London: Berg.

Quijano, Anibal. 2000. Coloniality of Power, Eurocentrism, and Latin America. *Nepantla: Views from the South* 1: 533–579.

Raeymaekers, Timothy. 2014. Introduction. In Id. (ed.). *ACME, An International E-journal for Critical Geographies* 13 (2):163–172.

Robinson, Cedric J. 1983/2005. *Black Marxism: The Making of the Black Radical Tradition*. Reprint. Chapel Hill: University of North Carolina Press.

Saucier, Khalil, and Tryon P. Woods. 2014. Ex Aqua: The Mediterranean Basin, Africans on the Move, and the Politics of Policing. *Theoria* 61 (141): 55–75. https://doi.org/10.3167/th.2014.6114104.

Small, Stephen. 2018. *20 Questions and Answers on Black Europe*. Amsterdam: Amrit Publishers.

Spillers, Hortense J. 1987. Mama's Baby, Papa's Maybe: An American Grammar Book. *Diacritics* 17 (2): 65–81.

Spivak, Gayatri Chakravorty. 1988. Can the Subaltern Speak? In *Marxism and the Interpretation of Culture*, ed. C. Nelson and L. Grossberg, 271–313. Basingstoke: Macmillan Education.

Stoler, Ann Laura. 2010. *Along the Archival Grain. Epistemic Anxieties and Colonial Common Sense*. Princeton: Princeton University Press.

Thompson, Robert Farris. 1984/2010. *Flash of the Spirit: African & Afro-American Art & Philosophy*. Reprint. New York: Random House.

Viswewaran, Kamala. 2010. *Un/common Cultures: Racism and the Rearticulation of Cultural Difference*. Durham: Duke University Press.

Wekker, Gloria. 2016. *White Innocence: Paradoxes of Colonialism and Race*. Durham/London: Duke University Press.

Werner, Marion. 2011. Coloniality and the Contours of Global Production in the Dominican Republic and Haiti. *Antipode* 43 (5): 1573–1597.

Wright, Michelle. 2015. *Physics of Blackness: Beyond the Middle Passage Epistemology*. Minneapolis: University of Minnesota Press.

Wynter, Sylvia. 2003. Unsettling the Coloniality of Being/Power/Truth/Freedom: Towards the Human, After Man, Its Overrepresentation—An Argument. *CR: The New Centennial Review* 3 (3): 257–337.

PART I

Borders

CHAPTER 3

When the Mediterranean "Became" Black: Diasporic Hopes and (Post)colonial Traumas

Angelica Pesarini

> I look forward to going to the government and to restore security throughout Italy, social justice and serenity. Anyone shooting is a delinquent, regardless of their skin colour but it is clear and evident that out-of-control immigration, an invasion like the one organised, wanted, and financed in these years, leads to social conflict. (Matteo Salvini's comments after the anti-Black shooting in Macerata on February 2018, Matteucci, 2018)[1]

> This is what we've come to in Italy: a climate of racist aggression is spreading, a racism that is directed not only against migrants but against anyone who does not have white skin, even against children adopted by Italian families. When people speak in general terms of populism in relation to this government they risk obscuring truly alarming facts on the ground with abstract political labels. There is no doubt that the blind eye this administra-

[1] Comments by Matteo Salvini on February 3, 2018 in relation to the anti-Blacks shooting committed in Macerata by former Lega candidate Luca Traini, in which six Black people were injured.

A. Pesarini (✉)
New York University, Florence, Italy
e-mail: angelica.pesarini@nyu.edu

G. Proglio et al. (eds.), *The Black Mediterranean*, Mediterranean Perspectives, https://doi.org/10.1007/978-3-030-51391-7_3

> tion turns to racist attitudes has had serious consequences. (Roberto Saviano, 2019)[2]

> Is there a racism emergency in Italy? Don't be stupid. I would like to remind that just in the last three days, in the general silence, the police have arrested 95 immigrants while 414 have been reported. Certainly the mass immigration allowed by the left hasn't helped matters. (Matteo Salvini's comments after the assault of Black Italian athlete Daisy Osakue, July 30, 2018, Martirano, 2018)

Roberto Saviano's impassioned piece titled "Wake up, Italians—our country is in a state of democratic emergency" that aimed to denounce anti-Black racism was published on *The Guardian* on March 2019. It marked the first-year anniversary of the Salvini – Di Maio government, the only European government led by two populist parties. It collapsed only fourteen months after its nomination.[3] Saviano's piece had been preceded by a vast amount of worry and concern for the Italian situation, including from the United Nations' Human Rights Chief, Michelle Bachelet, who, only six months after the Italian elections, announced her intention to send UN personnel to evaluate the gravity of anti-immigrant racism occurring in Italy.

In 2017, according to UNAR (the United Nations Anti-Racial Discrimination National Office), reports of ethnic-racial discrimination represented 82 percent of all complaints in Italy, while 2018 witnessed the increase of three hundred more cases in a year (UNAR 2019). The leader of the populist party *Lega*, the "captain" Matteo Salvini, as his followers like to call him, is considered by many as the proponent and the main trigger of the alleged "new" wave of racial hatred and anti-immigrant vitriolic rhetoric currently affecting Italy. Undoubtedly, Salvini, who centred his

[2] Roberto Saviano is an Italian writer author of "Gomorrah," a novel published in 2006 denouncing the crime of the *Camorra* gangs in the outskirts of Naples. Following the publication, Saviano received a number of death threats form the *Casalesi* clan so the then-Minister of the Interior, Giuliano Amato, decided to assign him police protection. Being a harsh critic of the former government, in June 2018 Salvini threatened to remove the police escort to Saviano. See, for example: https://www.ft.com/content/ff7d82c2-755b-11e8-aa31-31da4279a601

[3] On August 8, 2019, Matteo Salvini called for snap elections due to "diverging vision" within the two coalition parties Lega led by Salvini and Five Star Movement led by Luigi Di Maio.

electoral campaign on ideas such as "invasion" and "ethnic substitution",[4] put race and the Mediterranean at the very core of his campaign. By threatening the closure of Italian ports vis-à-vis of rescue boats, criminalising those helping migrants from drowning, implementing laws reducing asylum rights in Italy, boosting funding for police operations, and including provisions that allow for the revocation of Italian citizenship from foreigners who pose a "security threat",[5] Salvini managed to gain the trust of a conspicuous number of Italian voters, including southerners, once despised and ferociously insulted by Salvini himself.[6] *Lega Nord*, in its origins a provincial and rather small party arguing for the separation of the industrious and efficient northern Italy from the "lazy" South, managed to gain an impressive number of votes in the south of the country.

Despite perhaps seemingly well-intentioned pieces as Saviano's, such alarmist and dichotomised visions are often misleading. Begging Italy to rediscover her "better self", Saviano lists in his article a number of incidents occurred in Salvini's Italy in the first months of 2019, such as:

> In Lecce province, a young boy from Sierra Leone was battered on the back with a chair as his assailants racially abused him and told him to "go home." In Rome, a 12-year-old Egyptian was verbally abused and beaten up so badly by a group of older boys that he ended up in hospital. A black brother and sister were pilloried by a schoolmaster in Foligno, in central Italy. Women of colour are more and more treated as if they were sex workers—and not only in the street but even in public offices. Many incidents go unreported, but that doesn't take away from the fact that what is happening in Italy is a sign of a descent into barbarism. (Saviano 2019)

Towards the end of his piece, Saviano asks the reader: "How can one get people to see they are the victims of an evil spell?". Yet, it is precisely by separating roles and responsibilities in such clear-cut and dichotomised manners—good versus evil, refugees versus nationals—that populism has reached such a prominent position in Italian politics. Racist incidents, as

[4] See, for example: https://www.repubblica.it/politica/2017/12/03/news/salvini_non_cerco_voti_naziskin_ma_non_c_e_alcun_ritorno_del_fascismo-182878696/

[5] Decree on Immigration and Security, 2018.

[6] See for instance: "25 anni di insulti leghisti. Che il Sud non dimentica" by Orrico Di Mauro, December 19, 2019 http://www.facemagazine.it/25-anni-di-insulti-leghisti-che-il-sud-non-dimentica/.

the ones listed by Saviano, are certainly not unique to the Salvini period.[7] Despite openly racist and xenophobic stances, what Salvini found was a fertile soil on which to plant his seeds of hate speech. Anti-Black racism in Italy, in fact, did not emerge in March 2018 with the election of Salvini as Vice Prime Minister. Attributing the so-called "emergency of racism" in Italy to a sole individual would not only assign that individual with extraordinary capabilities, it would be far from reality and overlook historical implications of Whiteness in connection with race, national identity, and the repercussions of such ideas on racialised subjects.

Rather than imagining dangerous magicians tricking Italians with racist anti-refugee spells, it is important, I believe, to focus instead on the liminality of Italians' racial identity, the resignification of the Mediterranean category as an essential tool to whiten *both* Italy and Italians, and to bring light to how the history of colonialism has impacted the construction of a national identity marked by normative Whiteness. To do so, I use the framework of the Black Mediterranean, intended as a racialised and historicised space deeply influenced by the European colonial expansion in Africa. In the case of Italy, the Black Mediterranean may be considered as a useful analytical device, on the one hand able to debunk hegemonic discursive paradigms on issues of race, identity, and nation; and on the other hand, to reveal the Black Mediterranean as the physical and symbolic site of a reservoir of diasporic memories and narratives, excluded by the normative Whiteness of mainstream processes of knowledge production, and able to contrast the weight of dominant and exclusionary ways of thinking. The four sections of this chapter will show how the resignification of the Mediterranean category and its implication with race and colonialism have been essential for the construction of Italian national identity marked by a fragile Whiteness. At the same time, the presence and the negotiations enacted by racialised subjects within the Black Mediterranean challenges exclusionary spaces marked by normative Whiteness and demonstrate the importance of diasporic narratives as a form of resistance and counter history.

The chapter starts by illustrating first the internal racialisation of southern Italians at the end of the nineteenth century through the use of

[7] See http://www.cronachediordinariorazzismo.org/english/ a database monitoring episodes of racial discrimination in Italy. On decrees targeting a specific ethnic group see Pesarini (2013) on the Security Package passed by Silvio Berlusconi in 2008 and its repercussion on Roma communities.

pseudo-scientific theories promoted by prominent scientists such as Cesare Lombroso. One may notice here the influences of scientific racism on the conceptualisation of southerners' moral and physical inferiority, similar to Africans. The advent of the Fascist regime marks a turning point for the manipulation of the Mediterranean category. It was through the revitalisation of the cult of Rome and the exaltation of the Mediterranean heritage that Mussolini managed to achieve a so-called "whitening project" (Giuliani 2013: 578) by locating Blackness and "otherness" outside the border of the Nation. At this stage, we also witness the creation of the Empire of *Africa Orientale Italiana* by the fascist regime in May 1936. It is precisely in face of Africans, Mussolini argued, that Italians became aware of their racial identity and their duty to preserve such a title of nobility, as argued in the *Manifesto degli Scienziati Razzisti* also known as *Manifesto della Razza* published in 1938. Yet, the enactment of anti-miscegenation racial laws did not prevent the birth of Black mixed-race Italians, called *meticci*,[8] who will migrate to Italy at the beginning of the 1970s with an Italian passport in their hands. The last section of the chapter will focus on these specific experiences of migration and the analysis will draw upon some of the interviews collected during a period of fieldwork in Eritrea and Italy.[9]

[8] *Meticci* is the masculine plural form of *meticcio* (masculine singular), *meticcia* is the feminine singular form of this noun. In Italian when the plural refers to two or more nouns of different genders, the masculine plural is used. Therefore the noun *meticci* in this article refers to both male and female mixed race Italians. As Barbara Tizard and Ann Phoenix remind us: "Groups define and redefine themselves, their sense of who they are culturally and politically as preferred terms change. Also, within a group one person may like a term which another may not. We have to constantly pay attention to changing definitions and to the reasons why they are changed. People need to discover themselves who they are and not have terms imposed on them." (2002: 12). Therefore it is important to notice that the term *meticci* is used by many "mixed race" Italians but it is despised by others. I use it here because it was part of the colonial language of the historical time taken into analysis and also because some respondents in my research used it to identify themselves.

[9] For my doctoral research, I conducted interviews with two generations of women born from a Black African mother and White Italian father in the former Italian colonies in East Africa. The first generation includes the respondents born from the beginning of the Italian colonisation in Eritrea (1890) going through the Fascist invasion of Ethiopia (1935) and the defeat of the fascist army by the British in 1941. The second-generation sample includes those born after the official end of colonialism and fascism in East Africa in 1941 until the early 1970s. Interviews were conducted in Eritrea and Italy.

"This Is Not Italy, This Is Africa." The Mediterranean Category in the Aftermath of Unification

The South was cast as a form of other world racially different, a space to be explored, penetrated, contained and colonised. (Pick 1986: 62)

At the moment of its unification as a nation-state in 1861, Italy lacked a collective sense of national identity owing to the persistent presence of different rulers in several parts of the country for several centuries. Numbers of kingdoms and states brought about a remarkable cultural, political, economic, and linguistic diversity and, in this regard, it is striking to note that at the time of the unification, only 2,5 per cent of the population spoke Italian (De Mauro, 1991)). In the aftermath of the unification, Massimo D'Azeglio, one of the founders of the country, noticed that in order for Italy to exist, it was necessary to make Italians.[10]

Thus, the unification of Italy *and* the Italians turned out to be a very complex process, not only because of difficulty in combining a profusion of heterogeneous realities, but also due to the presence of an alleged internal "alien" world within Italy, located in the south of the country. At the end of nineteenth century, southern Italy was considered an alien part of the country inhabited by people belonging to a different civilisation (Dickie 1999; Giuliani and Lombardi-Diop 2013; Giuliani 2013; Pick 1986). During his permanence in the south, the patriot Luigi Farini powerfully wrote about the southern regions and its inhabitants in a letter addressed to Cavour on October 1860: "My dear friend...such savagery! This is not Italy, this is Africa—when compared to these peasants, the Bedouins are models of civil virtue" (Moe, 1992: 64).

The connections with Africa were not a threat due only to Italy's obvious geographical proximity to the continent. They were also perceived as a threat to the visible legacy of a shared history that, according to some, had tainted southerners' blood with that of Africans and Arabs that, in the words of Aliza S. Wong (2006: 4), "had coloured their progeny." One of the leading figures who attempted to provide "scientific" evidence of southern difference, and therefore southerners' close ties to Africa, was

[10] As D'Azeglio wrote, "Bisogna far gli Italiani se si vuol avere l'Italia (…) una volta fatti, davvero allora l'Italia *farà da sé*]" (D'Azeglio, [1867] 1949: 480).

the criminologist Cesare Lombroso. Considered as the father of modern criminology, Lombroso was interested in evolutionary theories of racial development and he closely associated race to innate criminal behaviour. Through his vast collection of data gathered from a variety of sources including cadavers, he traced the genealogy of the so-called atavist "born criminal" (Lombroso 1876/2006) by associating moral values to phenotypic features, through a concept known as the "atavist theory". In this configuration, the idea of race played a central role. If criminality was inherited, as Lombroso argued, then the members of certain "races" were genetically predisposed to be criminal:

> Race shapes criminal organisations. Both Bedouins and Gypsies can be considered races of organised criminals. The same seems to apply to Negroes in the United States (...), and to Albanians, Greeks and sometimes the indigenous people of southern Italy (...) Questions of race lead to issues of heredity. (Lombroso 1876/2006: 90)

In light of this theory, southern Italians showed what Lombroso identified as traits of deviancy, infirmity, and criminality. For instance, he attributed Sicilians' alleged criminal tendency to their descendance from "rapacious Arab conquerors" and their "bad blood" lacking of Aryan genes, a genetic legacy of Arab domination on the island (ibid, 118). Although Lombroso acknowledged the impact of social factors that affected southern populations, such as poverty, lack of education, inefficient institutional action, and illiteracy (ibid, 18), Whiteness and Blackness were crucial markers of categorisation in his analysis of crime. In 1871 with his famous publication *L'uomo bianco e l'uomo di colore. Letture sull'origine e la varieta' delle razze umane*[11] ("The white man and the coloured man. Readings on the origins and the variety of human races"), Lombroso clearly formulated the differences between what he considered the two main "general races": the White and the Coloured. Blacks represented an inferior stage of human development, and the sign of such inferiority was located in the body: a lighter and less developed brain, dark skin tone, the texture of the hair, and some physical features being very similar to those of apes. "Negroes", among which he included Africans, Australian Aboriginals, and African-Americans, represented a sort of living fossils from which derived the members of White races

[11] The first edition was published in 1871, followed by a second revised edition in 1892.

(Lombroso 1892/2004: 66). Yet, owing to certain genetic regressions, some individuals among Whites also embodied signs of Blackness that caused deviant behaviours, particularly people who had committed crimes, those affected by some mental health issues and individual born with specific conditions such as Congenital Iodine Deficiency, known at that time as "cretinism".[12] In regard of the latter, Lombroso noticed some common "monstrous" anomalies with southern *briganti*[13] and Blacks, namely that:

> At the beginning or at the worsening of their sad infirmities many alienated become darker and even their skin becomes bronze-like; and I saw in our valleys cretins who given the length of their skull, the protrusion of their muzzle and the thickness of their lips and even the darkening of their skin seemed *Negroes badly whitewashed.* (Lombroso 1892/2004: 66. Emphasis added)

The insinuation that southerners could be badly "whitened" Blacks is a crucial point here, as it tells us about the tensions and fears of being located in a liminal and ambiguous racial position. What Lombroso concluded is that, despite that the White race had reached the most advanced evolutionary stage of human development, this did not guarantee Italians a safe and clear reassurance of their racial status. Blackness was an insidious danger for Italy, much more so than for other European countries, due to a long and shared common history with Africa. Thus, according to the metaphor used by the Sicilian anthropologist and Lombrosian follower, Afredo Niceforo, if Italy was to be considered as an organism, not all its parts had developed equally. Some, namely the areas of *Mezzogiorno* (southern Italy), were less evolved and civilised. In their customs, southerners showed "the substance, if not the form, of past centuries. They are less evolved and less civilised than the society to be found in Northern Italy" (Niceforo 1901: 4). Therefore, a few years after its unification, Italy faced what came to be defined as *La Questione Meridionale*, or the

[12] Cretinism, now known as Congenital Iodine Deficiency or Congenital Thyroid Deficiency, refers to severe hypothyroidism caused by a deficiency of thyroid hormones during foetal and early life that restricts physical growth and causes learning difficulties. See: Salisbury (2003).

[13] *Briganti* referred to southern rebels who wanted to keep South's independence and fought against the new Italian State founded in 1861. See on this topic: Dickie (1999); Pick (1986).

southern issue, namely a number of discourses including the social sciences, medicines, and the humanities that tried to address the socioeconomic and political divergences between northern and southern Italy (Greene 2012: 3). It is precisely at this time that Italy started, according to some, its own "internal colonisation," a process aimed to civilise Italy's internal diversity (Giuliani and Lombardi-Diop 2013: 27) and to homogenise those areas of the country considered more backward and primitive. Nonetheless, to fully understand the extent and the complexities embedded in the process of southerners' "whitening", it is necessary to illustrate the impact that debates on Aryanism had on the racial construction of Italianness.

Polluted Blood and the Mediterranean Race

By the end of nineteenth century, figures such as Max Müller, Arthur de Gobineau and Ernest Renan had become leading intellectuals in the development of debates on Aryanism. The elaboration of racial classification based on the Aryan genealogy brought about the identification and ranking of Europeans into several distinct racial groups such as white Aryans on the one hand, and non-White and non-Aryan Semites, Hamites, and Mongolians on the other (De Donno 2006: 396). Owing to a perceived liminal and ambiguous racial positioning, Italy itself, and southern Italians in particular, became an object of discussion in relation to a sort of "polluted" blood that prevented them from being fully acknowledged within the Aryans. Of particular relevance for Italy was the work of the French aristocrat Arthur de Gobineau, who devoted a considerable amount of time defining the physical and moral features of "northern Aryans" in contrast to the degeneration of the "Mediterraneans" (Gillette 2002: 13–14).

Gobineau distinguished northern white Aryans from non-White southern Mediterraneans, specifically Latins and Semites (De Donno 2006). According to Gobineau, Rome was considered only partly Aryan given that the *alluvions ethniques* (ethnic floods) that had invaded her, caused the collapse of the Empire and a degradation of the Latin race still visible in the southern part of the country. In the second volume of "*Essai sur l'inégalité des races humaines*," Gobineau (1853) focused on the degeneration of Rome and the contamination of the Aryan blood with Semitic elements:

> It is time now to admire Rome through the new nationality that the ethnic floods gave her. Let's see what became of her when a more and more mixed bloodline imprinted on her, through a new character, a new definition. (Gobineau 1853: 249)[14]

Despite the "contamination", Gobineau admitted that Rome had been able to produce a significant number of artistic works, even though the authors of these productions were great spirits but not "real Romans" (ibid, 251). Due to the corruption of its blood, Roman civilisation fell apart and modern Italians could not be considered as the members of a pure race. Such a categorical exclusion of Italy from the Aryan genealogy was ascribed to southerners' different civilisation and their inferior racial stock, more similar to those of Semitic and Hamitic races (Pick 1986), namely Arabs and Africans. It is interesting to note how this same idea would be reiterated eighty years later by Hitler during an encounter with Mussolini in June 1934, in which he affirmed that Italian blood was mixed with "Negro blood" (Gillette 2002: 44).

The Sicilian anthropologist Giuseppe Sergi articulated an alternative answer to the so-called "Nordic racism" by formulating a racial classification who at the top was the so-called "Mediterranean race", namely Italics, the native peoples of Italy (Sorgòni 2002). In light with the "scientific" methods of his time, Sergi also conducted physical measurements to use as evidence for his claims. Despite collaborating with Lombroso, Sergi, unlike his colleague, considered skin colour as an irrelevant factor while the sole analysis of the skull was able to provide an effective human taxonomy (D'Agostino 2002: 323). It is by using cranial morphology that Sergi elaborated an Afrocentric theory in relation to race developments, published in English in 1901 and eloquently titled "The Mediterranean Race: A Study of the Origin of European Peoples." In the introduction of the book, Sergi affirmed that the early populations in Europe had originated in East Africa. From this "great African stock" (Sergi 1901: 14) he identified three varieties: the "African" who remained in the continent; the "Mediterranean" who settled around the Mediterranean Basin; and the "Nordic", who ended up moving to Northern Europe. These three variants belonged to what Sergi called the Eurafrican species. As a result,

[14] [Il s'agit maintenant de contempler Rome avec la nouvelle nationalité que les alluvions ethniques lui ont donnée. Voyons ce qu'elle devint quand un sang de plus en plus mêlé lui eut imprimé avec un nouveau caractère, une nouvelle précision.] My translation.

English, Germans, and Scandinavians were not Aryans, rather Eurafricans belonging to the Nordic variety, thus more related to Africans rather than the Aryans. Those erroneously known as Aryans, clarified Sergi, were in reality Euro-Asiatic groups that migrated from Asia to Europe in the Neolithic Age but were morphologically different from Eurafricans. They were rather a collection of people speaking the same Indo-European language. Sergi also argued that the European civilisation had to be traced back to the Mediterranean "race" since the Mediterraneans created the Minoan, the Hellenic, and the Latin civilisations that had spread all over Europe (Sergi 1901: 16). Therefore, two major claims were made: firstly, the cradle of Europe was in Africa and secondly, there was no racial differentiation between Africans and Europeans. Although his ideas on the Mediterranean "race" would be central in late Fascist conceptualisations, his theorisations about Hamitic people and Europeans belonging to the same species came across as problematic. As anthropologist Barbara Sorgòni (2002) notes, this implicitly put colonisers and colonised on the same level, precisely when Italy started its colonial invasion in East Africa. Thus, in order to establish a clear separation, the first legal colonial publications dating from the beginning of twentieth century marked a formal juridical separation from Italians and natives, in which these latter were considered "less civilised" and not "assimilable" in the realm of the nation (Sorgòni 2002: 44). These ideas will be furthered developed by Fascism through the assimilation of the internal Other, namely southern Italians, and the configuration of the new external Other: Africans.

Africa, Italy, and Fascism: The Whitening of Italians and the Resignification of the Mediterranean Category

> Naturally, when a people becomes conscious of its own racial identity, it does so in relation to all the races, not of one alone. We became racially conscious only in the face of the Hamites, that is to say the Africans... This is why the racial laws of the empire will be rigorously observed and that all who sin against them will be expelled, punished, imprisoned. Because for the empire to be preserved the natives must be clearly and forcefully aware of our superiority. (Mussolini's speech at the National Council of the Fascist Party, October 25, 1938)

The advent of Fascism (1922) played a crucial role not only for the homogenisation of all Italians within the realm of the Nation, but also for the construction of a racialised national identity based on the discursive production of the colonised Black "Other," and translated into the juridical discourse (Giuliani and Lombardi-Diop 2013). It has been pointed out by several scholars (Labanca 2002; Ben-Ghiat 2001) that the so-called Fascist "reconstruction" enacted by Mussolini concerned more the reconstruction of Italians rather than Italy per se. As Derek Duncan observes, according to Mussolini's project, Italians had to change not only their body but also their soul (Duncan 2005: 99). To do so, the regime organised the "reconstruction" following some imperative ideologies such as ruralism, anti-urbanism, anti-intellectualism, anti-bourgeoisism, antifeminism, pronatalism and homophobia (Bellassai 2005: 314–355). The aim was to build up a new Italian "race" and new Italian Fascists, aware of their identity and their Roman heritage. In order to provide all Italians with a sense of national belonging, the Fascist regime operated to reconfigure the Mediterranean category within the sphere of Whiteness (Giuliani and Lombardi-Diop 2013).

In the early 1930s, fascist intellectuals such as Nicola Pende drew on Giuseppe Sergi's ideas of "Latin Mediterraneanism" to criticise the idea of a pure Nordic Aryan "race" and to reclaim a Mediterranean "regeneration" (De Donno 2006: 401). The idea of "ethnic floods" mentioned earlier and identified by Gobineau as one of the main causes for the collapse of the Roman Empire, became the core of Latin Mediterraneanism. According to Pende, owing to the mixing of different "Mediterranean stocks" and concepts such as assimilation, unification, and universalism that the Roman Empire could maintain its greatness for several centuries and be considered the cradle of European civilisation (ibid, 402). Therefore, the so-called modern *razza Italica* (Italic race) identified by Pende was a Mediterranean race, originated from a number of thousand-year-old mixing of racial kinships, or *stirpi*, descending from the Romans and still having imperial blood running in their veins (Giuliani 2013: 579).

In this regard, Gaia Giuliani (2013) addresses the discursive manipulation of the Mediterranean category used by the fascist regime in order to enact a "whitening project" (Giuliani 2013: 578). By doing so, the regime managed to absorb the dark, southern, and internal "Other" within the realm of the nation on the one hand; on the other, it located Blackness and backwardness outside the borders of the nation (Giuliani and Lombardi-Diop 2013: 21–66). This discursive reconfiguration of the Mediterranean category became functional not only to separate Italian Mediterraneans from the colonised African counterpart, but also to enclose the alleged racial differences that had been negatively used to question Italians' purity into the Mediterranean racial identity (ibid). As Giuliani notes, it is through the appropriation and manipulation of the Mediterranean category that Italians managed to gain their Whiteness without clearly naming it. Rather than using biological evidence and colour, culture, and glorification of the past (the cult of Rome, or *romanità,* in particular) became the tools that helped the regime to "whiten" Italians (Giuliani 2013: 584). In line with such developments, it is interesting to note how, in 1932, Mussolini affirmed that race was merely a "feeling," and that no race could claim to be biologically purer than another (Gillette 2002: 44).

The regime's revitalisation of *romanità* represents what Ashley Montagu calls the "functional" role of the myth in conjunction with the legal discourse, namely the use of legal fictions to justify "the *status quo* endowing the myth with an aura of historical sanctity" (Montagu 1997: 42). The emphasis on *romanità* worked to provide the myth with historical sacredness, and to functionally render the myth of origins and the grandeur of Rome intelligible to Italians. *Romanità* was of great importance in reinstating in Italians a sense of pride in their ancestors and in shaping the idea of the new Fascist man so greatly desired by Mussolini. In this context, the legal discourse would be crucial for the materialisation of the racial "Other" no more located within the nation.[15]

The turning point for a shift towards new conceptualisations on race and identity can be traced to 1936, when Italy allied with Nazi Germany and announced the Empire of *Africa Orientale Italiana*. The war to annex Ethiopia, started in 1935, was particularly difficult for the Italian troops due to the intense resistance showed by the Ethiopian patriots. Italy won the war not only because of superior military equipment, but also because of the intensive use of poison gas (Del Boca 1976, 2005; Sbacchi 1997; Labanca 2002). After the proclamation of the Empire, the Fascist regime started to develop and reinforce segregationist policies aimed at a clear separation between Italians and Africans and to integrate the Nordic Aryan myth—criticised just a few years earlier—with the myth of *romanità* (Gillette 2002: 55). As Sòrgoni underlines, the aim of the racial scientists was to invent an Italian Mediterranean "race", distinct from the Germans but still traceable to the Aryan genealogy (Sorgòni 2002: 47). Owing to

[15] On the materialisation of the racial body in fascist East Africa see: Pesarini (2017).

new conceptualisations on "race" and blood, life in the colonies, and in particular the issue of *meticciato* (racial mixing) and interracial relationships, became a matter of great importance. Clear-cut legal regulations were essential, as Mussolini explained:

> The racial problem did not suddenly burst out as those who are habituated to brusque awakenings–because they are used to long armchair naps–would believe. It is in relation to the conquest of empire; since history teaches us that the empires are conquered with armies, but are held by prestige. And for prestige it is necessary to have a clear, severe racial consciousness, that establishes not only the differences, but also the clear superiority [of the imperial race]. (Mussolini, "Discorso di Trieste" September 15, 1938)

As a result of new discursive manoeuvres on race, we can see the production of the racialised colonial Other through the use of a series of racial laws and further segregationist policies implemented in the colonies in order to solve the so-called "racial problem" (Sorgòni 2002: 47).

Racial Laws and Refugees

While the regime enacted the first racial laws to avoid any contact between colonisers and colonised, interracial relationships had long started in East Africa, and several generations of Black "mixed race"[16] Italians, commonly named *meticci,* had already been born. Not surprisingly, after the declaration of the Empire in May 1936, the Fascist government started to immediately legislate against miscegenation, and the years going from 1937 to 1940 were, juridically speaking, quite hectic.

The first Fascist racial law, dating to 1937, prohibited relations of conjugal nature between an Italian citizen and a colonial subject, punishable with one to five years of imprisonment.[17] Meanwhile, 1938 saw the publication of the *Manifesto della Razza* (Race Manifesto) a document explaining in ten chapters the new biological stances on race adopted by the regime and the institutionalisation of racism as state ideology. In 1939 the regime passed the "Defence of Racial Prestige Against the Natives of Italian Africa" and introduced penal sanctions to punish the crimes

[16] "Mixed race" will be left in quotation marks to constantly remind the social construction of the term and to avoid reification.

[17] r.d.l. 19 April 1937, no. 880, turned into law 30 December 1937, no. 2590.

committed by Italians "in circumstances prejudicial to the race".[18] Finally, in 1940, a specific corpus of norms[19] stripped mixed race Italians of their right to gain Italian citizenship, denying blood affiliation with their Italian parent by declaring them as juridically "natives". As a result, it was prohibited for an Italian parent to legally acknowledge their children and provide for them.

Paradoxically, it is precisely at the pinnacle of racial segregation and physical separation in the colonies that the highest number of mixed-race children was registered. In 1931, a third of the children born in the colony were mixed (approximately 515 individuals), but in the three years of the racial law period, there were 2594 registered births and an unknown number of unrecognised children in Asmara alone (Barrera 1996: 43). Furthermore, through the massive presence of approximately 300,000 Italian soldiers temporarily staying in Eritrea due to the war against Ethiopia, the number of Eritrean women living with Italian men jumped from 10,000 in 1937 to 13,000 in 1939, and eventually rising to 15,000 in 1940.[20]

During my fieldwork in East Africa in 2012, I conducted interviews with two generations of women born during the racial laws period and in the aftermath of WWII to a White father and a Black mother (Pesarini 2015). The second-generation women in particular, born within the first two decades following the end of the war, were among the first Black Italians who crossed the Mediterranean in order to reach Italy due to the increasing levels of violence between Eritrea and Ethiopia. Having illegally annexed Eritrea in 1962, Haile Selassie, the Emperor of Ethiopia, was overthrown in 1974 by the Marxist DERG (Provisional Military Government of Socialist Ethiopia), guided from 1977 to 1991 by Mengistu Haile Mariam. A follower of Marxist-Leninist principles of nationalisation and land redistribution, Mengistu confiscated Italian properties and expelled all Italians in 1975. Meanwhile, the occupation of Ethiopia had given rise to two armed movements for Eritrean

[18] Law 1004/1939, articles 17 and 18.

[19] Norme relative ai Meticci, Law 822/1940.

[20] Four Power Commission of Investigation for the Former Italian Colonies, *Report on Eritrea*, London, 22 June 1948, app. 111, memorandum, in: Barrera, Dangerous Liaisons, 1996, p. 43.

independence: the ELF (Eritrean Liberation Front) and later the EPLF (Eritrean People's Liberation Front).[21]

As holders of Italian citizenship, many among my respondents could go back to Italy under the special status of *profughi*, namely refugees, protected by specific laws, such as the one enacted in 1952[22] that helped Italians coming from the former colonies after WWII. Unlike Eritreans, Ethiopians, and Somalis crossing today the water of the Mediterranean in very unsafe ways due to the impossibility of getting documentation, the trip of these Black Italians was very different. Despite fleeing a situation of war and violence, many among them were leaving on board military Italian ships from the coast of Eritrea to Naples and with an Italian surname written on their Italian passport. Nonetheless, their blood and name were not enough to make them Italians, as they realised once they set foot on Italian soil.

It is important to bear in mind that many Black mixed-race Italians loaded their migration process with high expectations, given that they were not just finding a refuge in a safer place. Rather, they were heading to their imagined and dreamt Italy, considered by some as a second motherland. One of the women who took part in my study, Rita, born in 1952 from an Italian fascist father and an Eritrean mother, explained to me how deep the process of Italianisation was for the recognised children of Italian men. In many cases this process brought about the complete erasure of Eritrean culture and language, as she experienced:

> At school we followed the Italian academic programme. The books and everything used to come from Italy. In fact, I don't know Eritrean history, as I never studied it (…) I really loved the *Divina Commedia* and I always used to say, "When I go to Italy, the first city I will visit is Florence!" (Interview, Asmara, 2012)

Rita expresses here her fantasies about Italy fostered by what she would learn at Italian schools in Eritrea. Interestingly, it emerges that fantasy played a crucial part not only for the construction of Africa by Italians but also for the construction of Italy by Eritreans[23] and Italo-Eritreans.

[21] See David Pool (2001).

[22] Law 137, March 4, 1952.

[23] For more on this topic, see Marchetti (2014).

Nevertheless, expectations and hopes seemed to crumble facing the impacts of the racialising practices operating in Italy and how the White gaze fixed and located the body of these individuals in specific racial positions of difference.

"We Are Not in Italy, Are We?" Postcolonial Traumas in the Black Mediterranean

When I mentioned to the respondent Bruna (born in Eritrea in 1940) that she was among the first "mixed race" Italians to arrive in Italy, she laughed and clarified to me they were not the first mixed Italians, rather "the first niggers." This episode speaks volumes about the painful identity reconfigurations and traumas these mixed Italians experienced. The "shock of the arrival" (Alexander 1996), namely the trauma experienced by many migrants in relation to the arrival in an unknown and unfamiliar place and the uncertainty about the possibilities of survival, added to a number of other specific traumas.

Talking with my respondents I noticed that besides the shock of the arrival, as theorised by Alexander, there were further shocking and traumatising experiences they had to face. One of these was what I call the "shock of the gaze". This refers to the visual shock caused by the clash between fantasy and reality: in this case the shock between Italy, as imagined in the colonies, and Italy (and Italians) in the 1970s. A perfect example is provided by Bruna who clearly describes the visual shock she experienced when her boat reached Naples:

> The impact with Italy was bad, shocking ... In Eritrea I would see wealthy and affluent Italians but when I arrived in Naples by boat, everyone was begging for money. It was a dirty city. Children were half naked. What a shock! (...) If you would see Asmara at that time, it was all flowers and gardens! And you could see all these women, with a white apron on, pushing the prams of Whites' children. They were all maids, our mothers. (Interview, Asmara, 2012)

Bruna clearly associated Italianness to wealthy economic conditions due to the fact that the majority of Italian settlers coming from the lower classes who went to East Africa had the chance to massively improve their economic status (Barrera et al. 2008) Seeing White Italians affected by poverty and deprivation represented the disruption of hierarchical views of race and class internalised by this respondent and the disruption of her fantasies embedded in colonial logics and based on her lived experience as a colonial subject. The subversion of such boundaries clearly caused a major breaking point in the societal paradigm many Black mixed-race Italians had internalised until that moment. Thus the "shock of the gaze" raised from the disruption of the what Hohmi Bhabha calls the "familiar alignment of [post]colonial subjects" (Bhabha 1994: 58) between the society in which they were born and raised and the reality of 1970s Italy.

Another great source of shock was caused at a linguistic level: on the one hand, the feeling of sudden linguistic disconnection in relation to local dialects never heard before and, on the other, the shock of seeing Italians being shocked themselves in front of a Black person speaking perfect Italian. Bruna recalled:

> Another thing really shocked me, and I would ask my husband [about it]: 'Hey, we are not in Italy are we? Because Italians speak Italian, right?' I couldn't understand the dialect, I couldn't at all! And I would ask my husband: "What language do these Italians speak?"...When I arrived in Italy they looked at me like if I was a freak because I spoke Italian. "How is that possible? You come from Africa and you speak such a good Italian!" I would sing *Sanremo*'s song![24] (Interview, Asmara, 2012)

The "shock of the language" was mentioned by several respondents, given that their Italianness was deeply forged on their linguistic identification. But the shock came not just as a confrontation to their mother tongue. These unpronounceable sounds from new dialects were also a shock to the tongue itself. Being an Italian native speaker was, in their view, one of the distinctive signs *making* them Italian, regardless of their colour. Once in Italy, however, they noticed with great surprise that their Italian linguistic and cultural capital was overlooked in favour of their colour. This leads to another trauma experienced by both mixed-race Italians but also those Eritreans, Ethiopians, and Somali who, for a number of reasons, had a connection with Italy before their migration. It is what I call the "shock of forgetfulness" because it originated from the

[24] The *Festival della Canzone Italiana di Sanremo* (Italian Song Festival of Sanremo) is held in the town of Sanremo, in the region of Liguria, and it is considered the most important and popular singing contest in Italy. Its first edition was broadcasted by RAI TV (the national public broadcasting company of Italy) in 1951.

"amnesia" of the colonial experience in East Africa enacted in the aftermath of WWII and the consequent erasure of the Black Italian experience.

The "amnesia" of the colonial experience refers to the selective forgetfulness of the colonial past and the concealment of Italy's responsibilities and consequent war crimes it committed while being an ally of Nazi Germany until 1943. In this regard, it is important to mention the historian Angelo Del Boca, who was among the very first who worked for the rehabilitation of historical integrity and the rejection of crystallised revisionist accounts of the colonial past. Through his archival research started in the 1960s, Del Boca brought to light several massacres committed by the Italians in Libya and East Africa,[25] and he was the one who discovered and denounced the use of mustard gas during the 1935 war against Ethiopia. Del Boca argues that the removal of the colonial past and the forgetfulness about the crimes committed by the Italians originated from a post-war lack of serious debate on the topic and a political intention to purposely pollute the "historical truth" (Del Boca 2002: 114). As Del Boca clearly highlights:

> This removal of the crimes is due primarily to the fact that in Italy, unlike other countries, a serious organic and comprehensive debate on the phenomenon of colonialism has never been promoted. It has been tried instead, by some state institutions, to muddy the waters with the precise design to prevent the truth to emerge, while a history of moderate or revisionist sign blatantly favours the removal of colonial guilt. (Del Boca 1991: 11. My translation.)

It is striking to note that despite the request of extradition of 1200 war criminals from the part of Ethiopia, former Yugoslavia, and Greece, the length of the trials was intentionally extremely long and these criminals never went to prison (Mellino 2006; Del Boca 2002). As a result, they were not only absolved, but even honoured by the institutions and were incorporated into the new post-war political parties (Del Boca 2002). The selective memory about the colonial experience also nourished the creation of romanticised myths about the Italian presence in Africa, the most famous one referring to the good heart of the Italian people, *italiani brave gente*. This expression alludes to the "generosity" of the Italians

[25] Between 1901 and 1932, Italians killed 100,000 people (soldiers, civilians and rebels) in Libya. In Ethiopia, they killed no fewer than 400,000 between 1887 and 1941 (Del Boca 2002: 113).

who, rather than benefitting from the colonial experience, unlike other European nations, built streets and hospitals and brought civilisation to East Africa.

Yet, breaking ties with the colonial past also meant the erasure of the Black Italian experience from the Italian collective imagination, as is evinced by those who arrived in Italy from the mid-sixties. Somali-Italian writer Shirin Ramzanali Fazel, who came to Italy in 1976, told me about this unsettling experience:

> I thought I knew Italy and Italians because I had always seen them in their environment as it was in Mogadishu at that time. But I ended up in a small province town where nobody knew where Somalia was. They would mistake it with Eritrea and Ethiopia, they would make it a whole: Africa. Or they would ask you odd questions like if we had cars in Somalia, or if we would wear grass-skirts, absurd things! Italians had no idea of what Africa was. To them it was ignorant and half naked people walking… They were surprised that I would speak Italian or some old man would tell me that in Somalia they had built streets and bridges and that they had taught us everything. Anything would amaze them: what I would eat, how would I speak, how would I walk…to me it was an incredible shock, I don't even know how to explain it. (Pesarini 2018)

The questioning of a person's "origins" or the ability of speaking a certain language is often used to interrogate individuals considered non-conforming to normative racial ideas of national identity, as has been investigated by a number of scholars (Mahtani 2002; Ali 2003; Haritaworn 2009). In this regard, language may be considered a tool able to reinforce structural relations of domination within hierarchal terms (Butler 1997). By asking questions such as "Why do you speak such good Italian?" we can see the power dynamics being involved in this interaction in which the listener, considered occupying a certain social position, may be injured by the utterance. In the case of the Black Italians interrogated about their ability to speak, one may notice the hierarchical relationship identified by Butler. The process of asking about ability of speaking Italian, despite hearing a perfect Italian accent, may be interpreted as a way to use the language in order to "act upon" the racialised subject and to deem her outside the symbolic national space. This process triggers what Butler (1997: 12) defines as "speech that wounds," given the unbalanced power involved within such an interaction. Nevertheless, Italian is precisely the tool used by these same subjects to respond to such potentially injurious utterances. By responding in Italian, they reinforce African writer Chinua

Achebe's powerful assertion: "For me there is no other choice. I have been given this language and I intend to use it" (Achebe, 1965:29). It is from within the Italian language, often imposed on Black mixed-race children to the detriment of their mother's local language, that subversion and resistance can be expressed. Their perfect mastery of the language, in some cases even more accurate than their interlocutor, allows them to respond and reclaim their own positioning. An interesting example is provided by respondent Saba (born in Somalia from an Italian father and an Ethiopian mother in 1968), who is well aware of her high level of education and mastery of the Italian language. She openly uses this asset to her own advantage in order to reposition unbalanced dynamics of power triggered by similar questions:

> My linguistic affiliation is very Italian and the fact that I speak Italian better than many Italians is something that I just show off against those who say, "Oh, but you are ..." or, "But how come you speak Italian so well?" I don't just speak Italian *so well*, I can teach you Italian *as well*, so shut up! [She laughs]. I am more Italian than them in this sense. (Interview, Rome, 2012)

These extracts show how certain questions imply an immediate repositioning of the Black Italian as not belonging and outside the nation. Certainly, they speak Italian well, or even *so well*, but that perfect mastery of the language is not enough to provide these subjects with membership to Italianness. Nonetheless, it is through the "speaking back" by using Italian that these same subjects may negotiate and reclaim their own positioning.

Conclusions: Diasporic Memories in the Black Mediterranean

Starting from the illustration of the daily news announcement of "alert racism" occurring in Italy due to the 2018 government election and adding to the obsession and spectacularisation of the "refugee crisis", this chapter has demonstrated how anti-Black racism in Italy does not trace its origins to the success of populist parties. Rather, it shows how current dynamics of anti-Black racism have been developing since the mid-nineteenth century in conjunction with Italians' fragile and liminal racial status. In this regard, the reconfiguration and manipulation of the Mediterranean category have been pivotal tools able to relocate Italians within the marker of European Whiteness and to distinguish them from African "otherness". Thus, colonialism in East Africa has been an essential tool in order to whitewash Italian national identity. It is precisely through the location of primitivism and inferiority in Africa that the fascist regime managed to absorb southern Italians within the realm of the nation. At the same time, the colonial invasion created the first large number of Black Italians who, later on, went to Italy to exercise their rights as citizens. Here my analysis has shown how categories such as "identity" and "belonging" may become exclusionary and traumatising mechanisms for those considered as "nonbelonging" and "out of place" (Puwar 2004; Mathani 2001; Root 1992), such as these first Black Italians who crossed the Mediterranean Sea.

The Black Mediterranean identifies, therefore, the malleability of social spaces within patterns of mobility and the importance of diasporic narratives as a form of resistance and counter-history. Despite certain bodies being framed as the "non-universal somatic norm" and dissonant due to their "invasion" of a space they are not considered to belong to (Puwar 2004: 8–52); the negotiations of racialised subjects occurring within this racialised and historically charged space, Black Mediterranean, tell us that that hegemonic paradigms of belongingness and national identity may be challenged, contested, and subverted, leading to alternative reconfigurations and (re)positionings of identity.

References

Achebe, Chinua 1965. " English and the African Writer." *Transition*, 4 (18): 27– 30.

Alexander, Meena. 1996. *The Shock of Arrival: Reflections on Postcolonial Experience*. Boston: South End Press.

Ali, S. 2003. *Mixed-Race, Post-Race: Gender, New Ethnicities, and Cultural Practices*. Oxford: Berg.

Barrera, G. 1996. *Dangerous Liaisons: Colonial Concubinage in Eritrea (1890–1941)*. Program of African Studies Working Papers [Online].

Barrera, G., et al. 2008. *Asmara. Architettura e pianificazione urbana nei fondi dell' ISiao*. Roma: Istituto Italiano per l'Africa e l'Oriente.

Bellassai, A. 2005. The Masculine Mystique: Antimodernism and Virility in Fascist Italy. *Journal of Modern Italian Studies* 10 (2): 314–335.

Ben-Ghiat, R. 2001. *Fascist Modernities: Italy, 1922–1945*. Berkeley/London: University of California Press.

Bhabha, H.K. 1994. *The Location of Culture*. London/New York: Routledge.

Butler, J. 1997. *Excitable Speech: A Politics of the Performative.* New York: Routledge.

D'Agostino, P. 2002. Craniums, Criminals, and the 'Cursed Race': Italian Anthropology in American Racial Thought, 1861–1924. *Comparative Studies in Society and History* 44 (2): 319–343.

De Mauro, T. 1991. *Storia linguistica dell'Italia unita*, Bari-Roma, Laterza

De Gobineau, A. 1853. *Essai sur l'inegalité des races humaines*, vol 2, Paris: Didot Frères.

Del Boca, A. 1976. *I gas di Mussolini.* Bari: Editori Laterza.

———. 1991. *Gli italiani in Libia. Dal fascismo a Gheddafi.* Bari: Laterza.

———. 2002. *L'Africa nella coscienza degli italiani: miti, memorie, errori, sconfitte.* Milano: Mondadori.

———. 2005. *Italiani brava gente? Un mito duro a morire.* Vicenza: Pozza.

Dickie, J. 1999. *Darkest Italy: The Nation and Stereotypes of the Mezzogiorno, 1860–1900.* New York: St. Martin's Press.

Donno, F. 2006. La Razza Ario-Mediterranea. Ideas of Race and Citizenship in Colonial and Fascist Italy, 1885–1941. *Interventions* 8: 394–412.

Duncan, D. and Andall, J. (esd.) 2005. *Italian Colonialism. Legacy and Memory.* Oxford: Peter Lang.

Gibson, M., and N. Nicole Hahn Rafter. 2006. Editors' Introduction. In *Criminal Man.* Durham: Duke University Press.

Gillette, A. 2002. *Racial Theories in Fascist Italy.* Psychology Press. London.

Giuliani, G. 2013. L'Italiano Negro. *Interventions: International Journal of Postcolonial Studies* 16 (4): 572–587.

Giuliani, G., and C. Lombardi-Diop. 2013. *Bianco e Nero: Storia dell'identità razziale degli italiani.* Milano: Mondadori.

Greene, S. 2012. *Equivocal Subjects: Between Italy and Africa – Constructions of Racial and National Identity in the Italian Cinema.* London/New York: Continuum Press.

Haritaworn, J. 2009. "Caucasian and Thai Make a Good Mix": Gender, Ambivalence and the 'Mixed-Race' Body. *European Journal of Cultural Studies* 12: 59–78.

Labanca, N. 2002. *Oltremare. Storia dell'espansione coloniale italiana.* Bologna: Il Mulino.

Lombroso, C. 1876/2006. *Uomo delinquente. Criminal man.* Trans. M. Gibson and N. Hahn Rafter. Reprint. Durham: Duke University Press.

———. 1892/2004. *L'uomo bianco e l'uomo di colore. Letture sull'origine e la varieta' delle razze umane.* https://www.liberliber.it/online/

Loomba, A. 2005. *Colonialism/Postcolonialism.* London: Routledge.

Mahtani, M. 2002. What's in a Name? Exploring the Employment of 'Mixed Race' as an Identification. *Ethnicities* 2: 469–490.

Marchetti, S. 2014. *Black Girls: Migrant, Domestic Workers and Colonial Legacies.* Leiden: Brill.

Martirano, D. 2018. Aggredita l'atleta azzurra Daisy Osakue, Salvini esclude il razzismo (ma Zaia no), *Corriere.it, July* 31 https://www.corriere.it/politica/18_luglio_31/aggredita-atleta-azzurra-polemica-074fe652-942e-11e8-827e-24bcbc32092b.shtml

Mathani, M. 2001. "I'm a Blonde-Haired, Blue-Eyed Black Girl": Mapping Mobile Paradoxical Spaces Among Multiethnic Women in Toronto, Canada'. In *Rethinking 'Mixed Race'*, ed. D. Parker and M. Song. London: Pluto Press.

Matteucci, Piera, 2018. "Raid razzista a Macerata, Salvini: "Colpa di chi ci riempie di clandestini". Renzi: "Ora calma e responsabilità", La repubblica, February 32018 https://www.repubblica.it/politica/2018/02/03/news/raid_razziale_a_macerata_salvini_chiunque_spari_e_un_delinquente_-187953469/

Mellino, M. 2006. Italy and Postcolonial Studies. *Interventions* 8 (3): 461–447.

Moe, N. 1992. "Altro che Italia!". Il Sud dei piemontesi (1860–61). Meridiana, 15, 53–89

Montagu, A. 1997. *Man's Most Dangerous Myth. The Fallacy of Race*. Oxford: Rowan and Littlefield Publishers.

Niceforo, A. 1901. *Italiani del nord e italiani del sud*, Torino: Fratelli Bocca Editori.

Pesarini, A. 2013. The Reinvention of Tradition: New Configurations of Gender Identity and Economic Strategies Within Roma Communities in Italy, *Refugee Survey Quarterly* 32(1): 101–121.

———. 2015. *Colour Strategies. Negotiations of Black Mixed Race Women's Identities in Colonial and Postcolonial Italy.* PhD thesis, University of Leeds.

———. 2017. "Blood Is Thicker Than Water". The Materialization of the Racial Body in Fascist East Africa. In "Performing Race". *Zapruder World* 4. http://zapruderworld.org/journal/archive/volume-4/blood-is-thicker-than-water-the-materialization-of-the-racial-body-in-fascist-east-africa/

———. 2018. Dinamiche neocoloniali di genere, "razza" e migrazione. L'universo femminile di Shirin Ramzanali Fazel in Nuvole sull Equatore. In *Donne e Sud*, ed. Ramona Omnis Darcel and Manuela Spinelli. Firenze: Franco Cesati.

Pick, D. 1986. The Faces of Anarchy, Lombroso and the Politics of Criminal Science in Post Unification Italy'. *History Workshop* 21: 60–86.

Pool, D. 2001. *From Guerrillas to Government: The Eritrean People's Liberation Front*, East African Studies. Athens: Ohio University Press.

Portelli, A. 2005. The Problem of the Color-Blind: Notes on the Discourse on Race in Italy. In *Race and Nation: Ethnic Systems in the Modern World*, ed. P. Spickard. New York: Routledge.

Puwar, N. 2004. *Space Invaders: Race, Gender and Bodies Out of Place*, Oxford: Berg.

Root, M., ed. 1992. *Racially Mixed People in America*. Newbury Park: Sage.

Salisbury, S. 2003. Cretinism: The Past, Present and Future of Diagnosis and Cure. *Paediatrics Child Health* 8 (2): 105–106.

Saviano, R. 2019. Wake Up, Italians – Our Country Is in a State of Democratic Emergency. *The Guardian.*

Sbacchi, A. 1997. *Legacy of Bitterness Ethiopia and Fascist Italy, 1935–1941.* Asmara: The Red Sea Press.

Sergi, G. 1901. The Mediterranean Race: A Study of the Origins of European Peoples. https://archive.org/details/mediterraneanrac00serguoft/mode/2up/search/whiteness

Sorgòni, B. 2002. Racist Discourses and Practices in the Italian Empire under Fascism. In *The Politics of Recognizing Difference. Multiculturalism Italian-Style*, ed. R. Grillo and J. Prtt. Aldershot: Ashgate.

Tizard, B., and A. Phoenix. 2002. *Black, White or Mixed Race?: Race and Racism in the Lives of Young People of Mixed Parentage.* London: Routledge.

UNAR 2019. https://www.open.online/2019/03/18/lallarme-aumentano-razzismo-e-xenofobia-lunar-trecento-casi-in-piu-in-un-anno-9-al-giorno/

Wong, A.S. 2006. *Race and the Nation in Liberal Italy, 1861–1911: Meridionalism, Empire, and Diaspora*, Italian and Italian American Studies. 1st ed. New York: Palgrave Macmillan.

CHAPTER 4

Fanon in the Black Mediterranean

On Violence, History, and Subjectivity

Gabriele Proglio

> From all these continents, under whose eyes Europe today raises up her tower of opulence, there has flowed out for centuries toward that same Europe diamonds and oil, silk and cotton, wood and exotic products. Europe is literally the creation of the Third World. The wealth which smothers her is that which was stolen from the underdeveloped peoples. (Frantz Fanon, *The Wretched of the Earth*, 1961: 102)

> My body was given back to me sprawled out, distorted, recolored, clad in mourning in that white winter day. The Negro is an animal, the Negro is bad, the Negro is mean, the Negro is ugly; look, a nigger, it's cold, the nigger is shivering, the nigger is shivering because he is cold, the little boy is trembling because he is afraid of the nigger, the nigger is shivering with cold, that cold that goes through your bones, the handsome little boy is trembling because he thinks that the nigger is quivering with rage, the little white boy throws himself into his mother's arms: Mama, the nigger's going to eat me up. (Frantz Fanon, *Black Skin White Masks*, 1952: 86)

G. Proglio (✉)
Centre for Social Studies, University of Coimbra, Coimbra, Portugal

G. Proglio et al. (eds.), *The Black Mediterranean*, Mediterranean Perspectives, https://doi.org/10.1007/978-3-030-51391-7_4

Introduction

According to an article published in *The Guardian* in June 2018, 34,361 people, trying to reach Europe, have died in the Mediterranean since 1993.[1] Alex Needham, the journalist who wrote this article, examined data supplied by UNITED, an association dedicated to documenting deaths that have occurred along numerous seashores of the world. The media in various countries generally use a specific vocabulary that includes words such as "emergency" and "tragedies" when talking about shipwrecks and deaths that have taken place in international and national waters, as if migrant arrivals were not foreseeable events. By contrast, many scholars have implicated European and North African institutions in this situation, clearly attributing a major portion of the responsibility for these deaths to them. For example, Danewid (2017) has offered important insights on the role played by Europe and the West in operations to rescue people from capsized boats in the Mediterranean.

My aim in this article is to apply Frantz Fanon's thoughts to problematize the condition of thousands of people at Europe's borders. In particular, as pointed out in the introduction to this volume, this contribution is positioned within the scientific debate on the Black Mediterranean. In the first section of this article, I introduce and explore Fanon's conception of violence. In the following section, I analyse symmetries relating to the conditions of natives in the colony and migrant people in the Mediterranean through references to Fanon's writings. The third section presents an in-depth analysis of the interlinkages between power, history, and subjectivity. In this final part of the article, the voice of Fanon is combined with those of migrant people whom I interviewed.

On Violence

Fanon (2008 [1952]: 5) wrote the following words: "Every human problem must be considered from the standpoint of time." This statement, which applies to the present time, indicates the need to contextualize every study within a historical moment. Thus, Fanon reveals his own subjectivity to readers through a "trick of mirrors" positioned between

[1] Alex Needham, The List: the 34,361 men, women and children who perished trying to reach Europe, *The Guardian*, 20 June 2018. The article can be accessed at: https://www.theguardian.com/world/2018/jun/20/the-list-34361-men-women-and-children-who-perished-trying-to-reach-europe-world-refugee-day

himself and the subjects of his investigation.[2] In my view, this method of situating himself within his time resonates with intersectional feminism, which affirms the positionality of the subject.

As Siebert (2012) has pointed out, Fanon's understanding of violence can be juxtaposed with that of Hanna Arendt, the German philosopher, for whom "violence, curiously enough, can destroy power" (Arendt 1958: 203). Another standpoint on the power–violence relation is that of Michel Foucault, who applied the notion of *dispositive,* referring to the mechanisms of power ([1977] 1980) and the production of genealogies. Writing from a different perspective, Fanon's interest lay in the act of violence embodied by colonized people. As Siebert (2012: 77) pointed out, the aggressiveness "is settled in muscles" and "tightens the body and mind and takes with them an ongoing tension." From this perspective, the effects of violence are manifested in psychological pathologies and a counter-violence that Siebert suggests is *negative* rather than *affirmative,* as argued by the French philosopher, Jean-Paul Sartre, in the introduction to *Les damnés de la terre* (1961). In that work, Sartre (1961: 20) stated the following: "for he shows clearly that this irrepressible violence is neither sound and fury, nor the resurrection of savage instincts, nor even the effect of resentment: it is man recreating himself." Evidently, Sartre's idea of recreating a unity recurs within an interpretative paradigm focusing once again on Europe. In fact, Fanon is more attentive to the effects than to the realization of a psychic tension in the reconstruction of the subject.

Hence, the first chapter of the *Wretched of the Earth* is dedicated to violence. Decolonization is a violent phenomenon that is aimed at changing the universal order and is considered a programme of absolute disorder (Fanon 1961: 34). Fanon states that the settler "brought the native into existence" and "perpetuates his existence"(35). The colonial world and its ordering and geographical layout remain important in marking out "the lines on which a decolonized society will be reorganized" (37). In this passage, Fanon affirms that colonialism constitutes an archive of forms of subjectification entailed in the reproduction of spaces for white and black people. Critiquing Marx's thought, he affirms that the colonial world is a Manichean world with effects on the psyche of colonized

[2] On this topic, see Achille Mbembe, "Préface," in Frantz Fanon, *Oeuvres*, La Découverte, Paris 2011: 9–21; Nigel Gibson, *Living Fanon. Global Perspectives*, Palgrave, New York 2011; Alice Cherki, *Frantz Fanon. Portrait*, Seuil, Paris 2000; David Macey, *Frantz Fanon, une vie*, La Découverte, Paris 2011.

people.[3] Specifically, "the native is declared insensible to ethics; he represents not only the absence of values, and in this sense he is the absolute evil"(41); he is able to destroy all that come near him as the depository of maleficent powers. Sometimes he is dehumanized or animalized (41), but he is always individualized by European culture. As Ania Loomba and Albert Memmi suggest, the "mark of the plural" is another process of de-subjunctivization at the hands of the white man (Loomba 1998; Memmi 1967). At the same time, specific features are attributed to a group marked by skin colour.[4]

Fanon states that "the settler-native relationship is a mass relationship. The settler pits brute force against the weight of numbers" (Fanon 1961: 52). This happen because the colonizer "is [an] exhibitionist," and "his preoccupation with security makes him remind the native out loud that there he alone is master" (52). The native, with a yearning gaze, hopes to occupy the settler's place of luxury. Fanon writes: "he expresses his dream of possession—all manner of possession: to sit at the settler's table, to sleep in the settler's bed, with his wife if possible. The colonized man is an envious man" (39). As a consequence of his specific positioning within society, his dreams "are always of muscular prowess; his dreams are of action and of aggression" (51). Thus, violence could be canalized in the battle for liberation, such as the one fought by the Algerian National Liberation Front.

The Wretched of the Mediterranean

Fanon's ideas are strongly relevant for a reading of the present. As pointed out by Iain Chambers, "wretched people belong to the denied and repressed past, the colonial one, and at the same time—as theorized by Fanon—[they] belong to our future" (Chambers 2013: 13). Drawing on Fanon's words, here I conceptualize the colonial world both as a border and as an archive to discern a shared geography of societies after decolonization based on the role of Europe and the West. I conceive of this world as a border upon which an imaginative idea of the Earth constructed through dichotomies (North/South; Europe/Africa; progress/barbary)

[3] In particular, Fanon analyses the effects of the relationship between white and black people in *Black Skins, White Masks* (see the bibliography).

[4] Fanon analyses all kinds of genealogies of power relations between white and black people in *Black Skin, White Masks*.

is laid out. At the same time, the colonial world constitutes an archive because colonial knowledge was produced through scientific disciplines, or "laboratories of modernity" (Mezzadra 2008) that were used and reused to develop ever new definitions and typologies of the Other (see Wekker 2016; Stoler 2008, 2016).

More generally, I would like to discuss the colonial word, considered as a recurrent image within postcolonial societies, both in former colonized countries and within Europe. Each image discloses—according to the positionality of the subject—an imaginary and a genealogy of power relations, practices, feelings, and psychic tensions, such as those concerning the black man, as pointed out by Fanon (1952). Each image reveals its imminence and belonging both to the past—it was generated during the period of colonial dominion—and to the present—because its meaning can be investigated in relation to the ideas of a society and of the Other. For Fanon, the colonized subject is an invention of European power. Derrida offers a similar conception of the figure of the migrant, who is the key for discovering the social structures and their workings in the present. Both the colonized subject and the migrant are empty images, devoid of a culture and with the ability to petrify individuals (Fanon 1961: 92; see also Ficek 2011), to spread anxieties and phobias[5] and to produce double absences (Sayad 1999).

Accordingly, I will apply Fanon's thought to the Mediterranean context which, in recent decades, has been the site of an ongoing migratory flow from Africa and Asia. The Mediterranean is constructed now, as it was during the periods of colonization of all of the European powers, as a border that functions to separate antinomic dichotomies (Chambers 2019; Proglio 2018a). Although the stories of migrant people were and continue to be silenced (Proglio 2018b) by a mainstream narrative that centres on political questions, it is possible to conduct an analysis of the violence imposed on these bodies that originates in the social conditions in the Mediterranean region.

The idea of a "crisis," for example, a "migrant crisis" or "refugee crisis," dominates the public discourse. In Latin, a "crisis" denotes a choice, a decision, or a decisive phase of an illness. By contrast, in Greek, the term means to distinguish or to judge. Thus, in light of the imaginary disseminated by the European media that has penetrated the public as well as

[5] For a discussion on this topic, see Nicel C. Gibson and Roberto Beneduce, *Frantz Fanon, Psychiatry and Politics*, Rowman and Littlefield International, London 2019.

private spheres, a "crisis" can be understood as an ongoing apocalypse entailing the sudden appearance of unwelcome bodies. A value must be assigned to these bodies as if they were standing in front of a court convened by white men, and a place must be found for them. The redefinition of Europe's whiteness is based on this process in relation to the skins of thousands of people.

As observed by Roberto Beneduce, these bodies mark "a threshold"; "a difference which cannot be ignored" (Beneduce 2013: 39). In fact, the black body signals the colonial past. New terms are used to assign positionalities to these subjects within European societies, such as "migrant," "refugee", and "asylum seeker." This taxonomy constitutes an irrefutable feature of images, as noted earlier, which are empty of meaning; the process of attributing meaning. Moreover, Fanon argues that a Marxist analysis of the economic sub- and superstructures remains outside of the chain of signification.

Furthermore, the term "migrant" alone does not refer to a specific type of person and could be used to describe European emigration on a global scale. If this term is qualified by adding the preposition "*in*" to "migrants in Europe", the scenario immediately changes. The imaginary of a white Europe populated by black people reveals a genealogy of relationships among racialized subjectivities. "You are rich because you are white, you are white because you are rich" wrote Fanon (1961: 39). This prophetic sentence remains tragically true in the present. This becomes apparent through an examination of internal processes occurring at the European frontiers. At these locations, skin colour is a crucial feature in the attribution of subjectivities and the consequent definition of forms of mobility between states. This is also evident in the numerous regimes of border control that have been documented and denounced by many associations, such as those implemented between Italy and France, Spain and France, Greece and Macedonia, Hungary and Romania and Serbia, Slovenia and Croatia, and Austria and Germany.[6]

Visuality constitutes one of the ways in which the border device works in the Mediterranean. Scholars have examined the spectacularization of shipwrecks and tragedies at sea (Cuttitta 2012), the role and many facets

[6] See Martina Tazzioli's work for a comprehensive analysis of what is currently happening. I especially recommend the following work: Marina Tazzioli, "The Temporal Borders of Asylum. Temporality of Control in the EU Border Regime," *Political Geography* No. 64 (2018): 13–22.

of visuality in arrivals (Giubilaro 2017), and the narrative of migrant illegality (De Genova 2013). This topic can be analysed, incorporating Fanon's attentiveness to the role of an incarnate aesthetic. Speaking about the colonial world, Fanon observed that European people "delimit physically, that is to say with the help of the army and the police force, the place of the native" (1961: 40). The totalitarian exploitation of these bodies entails a portrayal of a native civilization as "a society lacking in values" (40). The native is declared as being "insensible to ethics. He represents not only the absence of values, but also the negation of value. He is, let us dare to admit, the enemy of value and in this sense he is the absolute evil" (40). Re-reading these words in the context of the Mediterranean prompts a horrifying revelation: they fit this context perfectly. In fact, all of the contemporary voices articulating opposition to migration in Europe use the same narrative model. For example, the Hungarian prime minister explicitly stated that "we do not want our colour . . . to be mixed in with others."[7]

There are at least two possible ways of imposing a separation of black people from white people. The first entails a formal exclusion of black subjects from European societies with the effect of producing pockets of poverty in particular locations, such as the *banlieues* in France and suburbs in the main European cities, as well as informal and transit camps, such as those in Italy (Ventimiglia), Greece (Lesbos), Hungary (Subotica), and Macedonia (Lojane). This separation, perceived as urgent, has been institutionalized through the establishment of camps by various countries located along the European borders. Examples include those in Tarifa and Ceuta e Melilla (Spain) and in Gibraltar (UK) on the Moroccan–Spanish border; camps at Lampedusa, Pozzallo, and Caltanisetta (Italy), Tunis (Tunisia) and Birkirkara (Malta) along the central route; and in Lesbos, Samos, and Chios (Greece) along the Balkan route. This European obsession with dividing and avoiding contamination has proliferated on both sides of the Mediterranean Sea: detention and expulsion camps and refugee shelters established according to communitarian, international and national laws in Europe and the externalization of the European borders in North Africa (in Morocco, Tunisia, Libya, Sudan, and Turkey among other countries). The second possibility entails the differential inclusion of

[7] Victor Orbán's speech delivered on 8 February 2018. Please see the article on the UN High Commissioner for Human Rights, Zeid Ra'ad Al Hussein, at: https://www.ohchr.org/EN/NewsEvents/Pages/DisplayNews.aspx?NewsID=22765

Otherness. In Italy, for instance, the criminalization of illegal migration is simultaneously combined with the development of an illegal labour market in farmlands, such as those in Rosarno, Castelvolturno, Saluzzo, and Caserta to mention a few sites. Both devices, namely those of exclusion and of differential inclusion are calibrated according to a line of colour (Du Bois 1903) as well as other markers of subjectivization (e.g., gender, age, language, place of arrival etc.). More concisely, not only does Otherness occupy a unique positionality; the image of white Europe is simultaneously retained at the centre of the public imaginary through these multiple acts of separation.

At this point in time, the diplopic gaze, based on a chronological idea of time relating to the Other, known *yesterday* as a native and *today* as a migrant, becomes obsolete. What is signalled instead is a masculine anxiety concerning the domain. This image discloses the language, structure, and practices of power. Its emptiness can be attributed to its capacity for revelation: it is extremely dangerous because as Fanon has pointed out, it shows how "the settler's skin is not of any more value than a native's skin" (1961: 44). In contraposition to a Manichean vision, an understanding emerges that "the life, breath and beating heart are the same" (44). This discovery breaks up the previous order and causes anxiety for the native/migrant because "he is never sure whether or not he has crossed the frontier." This individual was and is "always presumed guilty." (44) This aspect seems irremediably connected with Michel Foucault's (1998) concept of biopolitics. In particular, this a priori guilt is "managed" at the Mediterranean border through the application of methods and tools for controlling access, such as digital fingerprint recognition using the Eurodac database (see Pugliese 2010). The digitalization of the *alterity trace* works from two different perspectives. On the one hand, it defines the possibility of mobility, that is, whether or not these people are allowed to cross borders. On the other hand, it assigns a place for these people within or outside Europe in accordance with the Dublin Regulation.[8]

What is at stake is whiteness, namely an idea of Europe that conquered lands and oceans and continues to impose a new order in countries located on other continents. This is the same concept of Europe that has been productive of many "discursive constructs" through knowledge production within various academic disciplines (Hall 1980, 1997); it is an entity

[8] This European law deals with the right to apply for asylum. According to the Dublin Regulation, individuals can only apply for asylum in the first European state of their arrival.

capable of naming and talking for the Other (Todorov 1982; Spivak 1988) and of absorbing worlds that exist outside of European borders, as argued by scholars like Edward Said (1979, 1993), Paul Gilroy (2004), and Hannah Arendt (1951). Thus, bodies are perceived as being marked by a trait for their identification: skin colour. Fanon (1961: 40) pointed to the warning implied by this perception, namely that these bodies are a "corrosive element, destroying all that come near [them]" (Fanon 1961: 40) and serve as "the depository of maleficent powers, the unconscious and irretrievable instrument of blind forces" (40). Moreover, they are animalized through the use of zoological terms, such as those used by the former Italian vice president of the Senate, Roberto Calderoli, who reproached his black colleague, Cécile Kyenge, the former Minister for Integration and a member of the Chamber of Deputies, whom he likened to an orang-utan.[9] Xenophobic European groups use a specific lexicon that includes words such as monkeys, black sheep, or rats, or evokes the idea of a "white race."[10]

Racism is not based on ignorance; on the contrary, as Fanon (1961: 41) observed, it is a capitalist tool used to bring people together to exploit them. In *Racism and Culture*, Fanon argued that the meaning of racism is mobile, or in the words of Stuart Hall (1997), it is a floating signifier. Specifically, Fanon (1964: 32 stated: "racism has not managed to harden. It has had to renew itself, to adapt itself, to change its appearance. It has had to undergo the fate of the cultural whole that informed it."[11] From this perspective, the upgrading of racial categories is an output of the work of the colonial archive: an ongoing regeneration of those archetypes elaborated within knowledge[12] generated during periods of dominion of European powers on other lands and bodies (Fanon 1959). One of the main properties of this archive entails its flexible adjustment of both the

[9] Calderoli was sentenced to 18 months of prison for insulting Kyenge, aggravated by the racial factor. See: https://www.ilfattoquotidiano.it/2019/01/14/kyenge-orango-calderoli-condannato-per-gli-insulti-allex-ministro-aggravante-razziale/4895789/

[10] Annamaria Rivera, "Il caso Fontana e il neorazzismo all'italiana," *MicroMega*, 20 January 2018. http://temi.repubblica.it/micromega-online/il-caso-fontana-e-il-neorazzismo-allitaliana/

[11] Frantz Fanon made this statement before the First Congress of Negro Writers and Artists in Paris in September 1956. The speech was published in the Special Issue of *Présence Africaine*, June–November 1956.

[12] For a discussion of this aspect, see Frantz Fanon, "Medicine and Colonialism," in *Studies in a Dying Colonialism* (New York: Grove Press, 1965).

signifier and the significance of new bodies and social contexts, such as those described in the above-mentioned cases.

The main question at the core of Fanon's writings seems to be that of the liberation of human beings. As previously stated, violence is considered unavoidable to reach this goal. There are two typologies of violence. The first is derived from European powers, and is therefore intrinsic to the colonial world, which has "brought the native into existence and perpetuates his existence" (Fanon 1961: 35).The second relates to natives who have become conscious of their situation. Of such natives, Fanon writes: "he is overpowered but not tamed; he is treated as an inferior but he is not convinced of his inferiority. He is patiently waiting until the settler is off his guard to fly at him" (Fanon 1961: 52). This passage seems apposite for describing present conditions in Europe. The image of "migrants" stems from a process of subjectification of people who arrived in Europe. This process, which has involved all fields of European knowledge production (universities, executive agencies, the judiciary, legislative bodies, the media etc.) is part of a war waged against those subjects phagocytized by the word "migrant." As during the colonial period, the fear of European institutions relates to violence; not the violence of "migrants" against their fellows, that is, the tribal fights that Fanon (1961) referred to but to violence against Europe's social order. At this point, as Fanon (1961: 57) observes, the native "discovers reality and transforms it into the pattern of this custom, into the practice of violence and into his plan for freedom."

Fanon conceptualizes and proposes a collective and organized fight to achieve Algeria's liberation from France. Clearly, I do not intend to draw a comparison between the National Liberation Front guerrillas of Algeria and the strikes and protests that have erupted in Europe in recent years. On the contrary, my intention is to focus attention on the typology of responses to counteract European powers. Fanon points to the formation of a collectivity as a mechanism that is able to overcome differences among people. In this sense, there is a degree of similitude with the current postcolonial condition. On many occasions, protests have centred on claiming rights. An example is Balzi Rossi's protest that spread along the marine border between France and Italy in June 2015 after the French government decided to close the frontier and suspend the Schengen Agreement relating to migrant people who wanted to apply for refugee asylum status. On that occasion, a group of Sudanese people along with Eritreans, Somalians, and some Italians decided to camp on the cliffs in close proximity to the border. They remained there and built up the "No borders" camp, which remained active for 100 days up to 30 September when the police decided to evict them.[13] Similar protests occurred in Ceuta and Melilla in June 2018, when many people tried, collectively, to cross borders illegally[14]; and in Athens, refugees occupied the City Plaza Hotel.[15]

The media has presented a xenophobic image of the "migrant" as a criminal, if not a terrorist, who is trying to cross the Mediterranean.[16] Arriving on boats in Lampedusa, Lesbos, or the Spanish seaside, such people are portrayed as having the goal of attacking and damaging European society. Drawing on just this kind of imaginary, Prime Minister Matteo Salvini, the leader of the racist Northern League Party, came up with his absurd and criminal idea of closing the harbours to boats operated by NGOs in the Mediterranean for rescuing people in danger.[17] At best, a "migrant" is depicted as a stealer of jobs; otherwise, the colonial image of a (male) migrant as a rapist of white women endures. Individuals labelled as migrants are seen to work in illegal trades, such as drugs, prostitution, robberies and extortion. When this male migrant's religious identity is the focus of attention, he is depicted as an Islamic extremist imposing Sharia law on his wife and forcing her to wear *burkha*. All of these different facets of the same image originate in stereotypes of natives, Muslims, or Arabs from the colonial past conjoined with a narrative highlighting the difficult conditions endured by European emigrants outside of Europe from the end of the twentieth century.[18]

The inferiorization of migrants is not limited to the dissemination of narratives and imaginaries in the public domain. As pointed out by Ziauddin Sardar in the Foreword to the 2008 edition of *Black Skins White Mask,* the black man "must be black in relation to the white man" (Fanon

[13] For further details on the No Borders Camp in Ventimiglia, see https://noborders20miglia.noblogs.org

[14] See Hannah Strange's writing published by *The Telegraph* on 26 July 2018, "Dozens injured as hundreds of migrants storm Spain's Ceuta border with Morocco." https://www.telegraph.co.uk/news/2018/07/26/dozens-injured-hundreds-migrants-storm-spains-ceuta-border-morocco/

[15] See Molly Crabapple, "This refugee squat represents the best and worst of humanity," *The Guardian*, 23 June 2017.

[16] See Mazzarra (2019) and Cuttitta (2012) in the bibliography.

[17] See the BBC report "Migrant crisis: Italy minister Salvini closes ports to NGO boats," 30 June 2018.

[18] On colonial stereotypes in the postcolonial context, see Gaia Giuliani, *Race, Nation and Gender in Modern Italy: Intersectional Representations in Visual Culture* (New York: Palgrave, 2018), which is the most complete investigation on this topic.

1952 [1988]). Fanon refers to this internalization process as one of *epidermalization*, in which the ego of the black man collapses when it comes into contact with the white world. Roberto Beneduce has written powerfully on this aspect of colonial contact, which reverberates with the postcolonial condition (Beneduce 2004, 2007, 2014). In this article, I would like to focus specifically on some questions concerning Italy and Tunisia. Europeans in Tunisia use the word *expat,* which is a slang term for expatriate, to describe their condition. Black people on the opposite shore of the Mediterranean, in Italy use other terms such as "2G," which means second generations,[19] or Black Italians, Afro-Italians and "Nappy Italians."[20] In all of these terminologies, "being Italian" is foregrounded and prioritized. To grasp these differences in terminology, it is essential to understand that Italian law centres on *Jus sanguinis*, the principle of nationality whereby Italian citizenship is conferred on an individual if at least one parent is Italian. Although an important campaign was launched in support of a proposed reform of this law, it was unsuccessful. In March 2018, the proposal of the "leftist" government led by Paolo Gentiloni to approve a "humanitarian mission" in Niger for stemming migratory flows was approved in Parliament.

Hence, it is necessary to situate this reflection within a social context in which there is opposition to the hosting of new subjectivities within the imagined community (Anderson 1983). Consequently, persistent attempts to affirm the presence of "migrant people" in Italy, which is clearly evident but unacknowledged at the institutional level, could be interpreted as both a symptom and an effect of the process of inferiorization. Why is it not possible for these people to define themselves simply as Italians if they so desire, or as "Europeans among other things" to borrow a phrase used by Jacques Derrida (1991: 34).

This approach of imposing boundaries of nationality that determine the legitimacy of individuals and groups within a community also has impacts within the cultural world. Indeed, the goal of some researchers has been to "discover" and reveal the presence of black people within European and national history, as such discoveries can change conditions in the present. However, I question the relevance of these investigations, and their

[19] Please see the second generations network at http://www.secondegenerazioni.it

[20] This movement is based on an aesthetic and cultural claiming of roots outside of Europe. It comprises a community of girls who decided to "flaunt their natural Afro curly hair" according to the website: http://www.nappytalia.it/afro-italian-nappy-girls/

political value. What is happening in Italy seems to me to be comparable to the history of feminism, that is, to the first wave of protests that were directed towards affirming the equality of men and women. At the same time, I would like to support each and every fight for the rights of individuals and groups: the fight should be for all. Thus, my reflections are not intended to be an accusation, or a judgment; rather, my aim is solely to present an analysis of the situation.

Fanon postulates the existence of a "zone of not being," namely "an extraordinarily sterile and arid region, an utterly naked declivity where an authentic upheaval can be born" (Fanon 1988: 2). He observes that such places are the products of Europe and are populated by several characters. All of them have been generated through necropolitics (Mbembe 2003). There are people in these places who have been animalized and forced into inhuman conditions but who have managed to survive. There are also the faceless dead, who lack a story and who were abandoned and set adrift in the temporality of history. There are also "middle-characters"—half-human and half-animal creatures suspended between life and death—that echo classical mythology. I am speaking here of zombies who are manifestations of the master/slave dialectic. Several intellectuals (e.g., Hegel, Depestre and Mbembe) have analysed this image in relation to the condition of slaves. In the postcolonial condition, this image is evoked by the arrivals of groups, for example, in Romero's movies[21] portraying boats on the European seaside and stems from an individually and collectively felt anxiety relating to the bodies on which white violence has been inflicted. These bodies are dangerous because being undead, they are able to attack, penetrating each and every household and reversing established power relations.

White History and Black Subjectivity

The question that is raised is: what kind of history is possible in order to describe such a violent Europe? I was haunted by this question, which continually echoed in my mind and soul while I was conducting fieldwork on the European borders.[22] I immediately understood the significance of

[21] See *Night of the Living Dead* (1968), one of several fantasy movies directed by George Andrew Romero.

[22] The title of my research project is 'Mobility of memory, memory of mobility: Western Mediterranean Crossings in the XX and XXI centuries". The Fundaçao para a Ciencia e a

these places. They were not just frontiers located between two nations; nor were they simply devices used to control mobility. These places had and still have an epistemological relevance for describing what contemporary Europe is. History has been framed and recorded around the idea of the border, beginning with a line termed as a frontier.

Ventimiglia, the last Italian town before the French border, can serve as a starting point for my reflections. In a small section of Ventimiglia known as Grimaldi, a path runs between the two national frontiers of Italy and France. Initially, this path was known as "Hope's Path" referring to the escape of Italian socialists and communists from Fascist Italy. Then, someone traveling along this difficult route died after falling off a cliff, and consequently the name was changed to "Death's Pass." All kinds of people fleeing from the tragedies of the nineteenth and twentieth centuries have travelled along this path that is just one metre in width and stretches over a distance of four kilometres. They include Jews, Kurds, Palestinians, former Yugoslavians, Albanians, Rumanians, Afghans and individuals fleeing from the socio-political instabilities of the Southern Sahara and Maghreb regions. They have all tread the same path, advancing along the same track, step by step.[23]

I was deeply affected on my first visit to Death's Pass. All around me, articles of clothing and personal items lay strewn within a radius of just one metre: a red t-shirt here, a pair of black pants there and a black toothbrush a few feet away from me. A grey backpack was covered over with earth and small stones, and there were two thermal blankets rolled together. Three burned down cigarettes and empty water bottles, the labels of which had been yellowed by the sunlight lay before me. I looked up, and what I had just seen so close to me was repeated 2, 5, 10, 15, 30 metres away from me, as far as I could see into the distance. Traces: these

Tecnologia has funded this research project which is based at the Centre for Social Studies, at the University of Coimbra (2017–2023). The aim of this project is to study human mobility and its memories in the Western Mediterranean area during the Twentieth and Twenty-first centuries; particularly, in both directions between Morocco and Spain, Algeria and France, and Tunisia and Italy. Please, visit the project's website: https://www.mobilityofmemory.com

[23] On the history of the Death's Pass, see *Enzo Barnabà, Il passo della morte. Storie e immagini di passaggio lungo la frontiera tra Italia e Francia*, Infinito Edizioni, Genova 2019. See also Il filo e le tracce—XXmiglia, a video produced by Martina Melilli and Caterina Erica Shanta, with images and text by Gabriele Proglio (2019). https://www.youtube.com/watch?v=FLs-P9KCfBI

were traces of a history that was impossible to narrate, revealing fragments of lives by disclosing dark abysses of memory along life trajectories.

From that moment onward, I felt two contradictory feelings arise and grow inside of me. On the one hand, I experienced a sense of powerlessness and despair concerning what had transpired on that path, and in other parts of Europe. On the other hand, I became aware of the privilege of writing history accorded to me—as a white Italian scholar. Like every codified language, history is based on roles, and especially on proofs and references to scientific debates. Could these traces and fragments be part of history? Or will stories of migration only be told through numbers—how many people arrived in Europe in a year—and institutional actions (of the jurisprudence, police, humanitarian aid etc.). I believe that it is crucial to return to Fanon's thoughts in order to reflect upon this tricky and pivotal question, which is directly connected to another key question, namely, who is the owner of history? Fanon (1961: 50) postulates that "the settler makes history," and his own life, considered as a part of a national project of progress, served as the metric of his narratives. He continues: "He [the settler] is conscious of making it" because as he clearly indicates, he himself is an extension of the mother country. "Thus, the history which he writes is not the history of the country which he plunders but the history of his own nation in regard to all that she skims off, all that she violates and starves" (50). The native is condemned to immobility in the face of history's making. He has just one solution: "to put an end to the history of colonization, the history of pillage—and to bring into existence the history of the nation, the history of decolonization" (50).

More generally, history is deployed to measure power relations such as in the case of the Arab world, drawing a metrical comparison in terms of time: before colonialism when "history . . . was dominated by barbarism" (Fanon 1961: 127). Fanon's vision of history is not only linear, entailing a chronological order. In fact, as he affirms, every analysis has to be located in its specific context of belongingness. But not just this; as previously argued, his attention is focused on the narrative structure as well as on its effects, that is, on those breaking points—in the past, present or concerning possible futures—that are not visible from the sweetened and institutionalized narratives of European nations. These breaking points, which crack in the articulation of "cause and effect" models, arise from those forgotten, invisibilized, and dehumanized bodies of yesterday and today. Thus, Fanon's humanism leads him to make an urgent statement: "It is a question of the Third World starting a new history of Man, a history which will have regard to the sometimes prodigious theses which Europe has put forward, but which will also not forget Europe's crimes" (Fanon 1961: 195).

In the opening lines of *Black Skin White Masks*, Fanon states that "the black man is not a man" (Fanon 1954: 1) and must be extricated from the core of a white universe. He continues: "The black man wants to be white. The white man slaves to reach a human level". The only way for the native to find meaning seems to be to dig "into [his] own flesh," transcending that skin into which stereotypes and epidermization have become crystallized within the melatonin. Fanon (1954: 8) explains that when cognizance of the situation concerning power relations is attained—both within the colony and in the postcolony—"what matters is not to know the world but to change it." However, he provides a warning regarding the narrative device used for knowledge production: to speak means to be in a position to use a certain syntax, to grasp the morphology of this or that language, but it means above all to assume a culture, to support the weight of a civilization" (Fanon 1954: 8). Fanon is not only talking about a literate culture, but also about an oral one. In general, in a process that Fanon (1954: 84) refers to as historicity, the black individual is inscribed within the white narrative.

Several informal camps[24] are operative along Europe's borders in the northern part of the Mediterranean region, for example, in Bihaic (Croatia), Ventimiglia (Italy), and Idomeni (Greece). These places reflect the need for a personal and collective space that has been denied by the concerned institutions. Let us now enter these camps. Tents are everywhere, and their organization conforms to the sites where they have been established, meaning that their placement has not altered the surrounding environment. The tents are positioned in line with each other. There are blankets here and there, scattered on the floor. Some of them cover the upper portions of the tents in an attempt to keep out the damp, at least partly. There is dirt everywhere: inside and outside the camp. Black t-shirts, packets of cigarettes, glass bottles, black socks, pieces of toilet paper, pieces of carrots; all of these items fit together within a huge *mélange*. There is a smell of burnt wood. Fires are lit at night to keep hundreds of people

[24] See the UNHCR website page on the situation of refugees, in which arrivals in Europe across the Mediterranean are mapped at: https://data2.unhcr.org/en/situations/mediterranean. The Migreurop website provides details on how many camps, of different types, are located in the Mediterranean region: www.migreurop.org

warm. The camp is organized in sections, each of which is occupied by people of a particular nationality: Sudanese, Eritreans, Somalis, and Afghans. In all of the camps that I visited, some of the nearby residents displayed racism against the "migrants." The violence not only emanates from outside of the camp; it also erupts within it: violence between "ethnic groups" and violence against women, some of whom are obliged to sell their bodies within the camp. In this difficult context, there is no institutional aid offered; the only assistance available to these people is provided by volunteers from different institutions and NGOs. To the contrary, actions have been taken against them, such as the introduction of local laws targeting "immigrants" as in Ventimiglia, where Mayor Enrico Ioculano, who belongs to the "leftist" Democratic Party, promulgated an ordinance prohibiting the provision of food and water to "migrants."[25] As Fanon pointed out, these people have been and continue to be dehumanized and animalized. They are subjected to various forms of violence inflicted on their bodies by maintaining silence or through self-inflicted injuries and attacking others within the camp.

In light of these traumatic emotional experiences, and in that face of a difficult question, as previously stated, that requires a response, it seems salient to revisit the historical debate regarding the category of the subaltern. The term "subaltern," which was coined by Antonio Gramsci in the context of a political programme to transform conditions in Southern Italy (Gramsci 1971), was reintroduced as an analytical concept by the Asian Subaltern Studies group in the Indian context (Guha 1982). As stated by Ranajit Guha, a member of this group, the idea of a "history from below," conceived as a history of subjects produced through subtractions, or "an identity-in-differential" (Guha 1982: vii) was at the heart of this project. In particular, the voice of Gayatri Chakravorty Spivak was prominent within the group's internal debates. Spivak criticized the group's approach, which, she argued, was marked by an essentialist deconstruction practice in relation to subalterns. The question posed in one of her most widely read articles titled "Can the subaltern speak?" (Spivak 1988) appears to be integral to the subject of the Mediterranean. Spivak argued that the use of the category "subaltern" entailed objectifying subjects invisibilized by history. Critically, she suggested that the subaltern's agency is always constituted by European powers or through nationalist narratives.

[25] See Thomas Mackinson, *Ventimiglia, vietato dare cibo e acqua ai migranti. Solo la mobilitazione costringe il sindaco Pd a revocare l'ordinanza*, Il fatto quotidiano, 23 April 2017.

Spivak (1999) further argued that the native woman is foreclosed, unable speak or be heard because there is always a man—a settler or a native—who speaks for her. This *epistemic violence* is produced by Western cultures. It is a consequence of narratives that support as well as contest power. Spivak contended that Western thought[26] reproduces itself in the same moment that it criticizes itself, and this foreclosure also applied to the Subaltern Studies Group, which did not recognize the impossibility of the native subject being able to represent himself or herself. As Marx wrote in *The Eighteenth Brumaire of Louis Bonaparte*, "he must be represented" (124). Thus, Spivak (1999: 265) argues that the individual subject is reintroduced through totalizing concepts of power and desire.[27]

Strategies for a Theoretical Engagement with the Black Mediterranean

In light of my comprehensive and in-depth engagement with the scholarship of Spivak and others,[28] I will now attempt to elaborate theoretical strategies that can be deployed to analyse hidden, invisibilized and silenced Mediterranean stories. However, before proceeding any further, I would like to make one thing clear: addressing this question of violence from an epistemological perspective constitutes the first turning point in the deconstruction of knowledge. This means seeing beyond "the colour black" of the Mediterranean and deconstructing but not essentializing it. It means recognizing from a perspective that is against the light that this is a sea whose history is full of dark spaces where the stories of thousands of people have been and continue to be silenced and invisibilized. From this perspective, the Black Mediterranean represents the dark side of European history that is past as well as in the making. It is one of the borders at which contradictions surface and spill over into the entire world, for example, migrations in Europe as consequences of neo-colonialism and European political interference in Africa.

[26] Spivak criticized the two philosophers, Michel Foucault and Gilles Deleuze, for their lack of consideration of the diversification of subjects in their analyses of power relations.

[27] See *Concerning Violence. Nine Scenes from the Anti-Imperialistic Self-Defence* produced by Annika Roggel and Tobias Hanson, with an introduction by Gayatri Chakravorty Spivak, in which Lauryn Hill narrates text by Frantz Fanon.

[28] In particular, I am thinking of Partha Chatterjee, Homi K. Bhabha, Dipesh Chakrabarty and Edward Said.

At the same time, when discussing the Black Mediterranean, I feel that it is important to move beyond European borders and to be aware that this history, forged through many individual and collective stories that began with the slave trade and extend up to recent migrations, does not only belong to Europe. It is also a history that is lived on the skins of millions of people, past and present. It is a history constituted in fragments that are scattered in various places across the world (Africa, America, Europe etc.) and across various times and temporalities.

Any and all attempts to attain this goal of producing another typology of history signify the practice of decolonizing history because the Black Mediterranean came into being at Europe's borders. I am referring to "borders" both in the material sense of frontiers, walls, camps, and mobility control as well as from an epistemological standpoint. In fact, this boundary can be conceived as the production of a narrative about the chronological order and the classification of subjects in accordance with a colonial archive. In other words, there is a *tangential line* drawn between Europe and a multitude of diasporic cultural universes in the construction of black people's identities.

Thus, the black body becomes an incarnated border and every narrative aimed at resisting the hegemonic power strategically exceeds the discourse order (Foucault 1971) and the representation of the other (Derrida 1991). This process is already underway in thousands of homes and communities across Europe. The above-mentioned line is constantly being crossed through practices and within oral cultures as well as group structures and organizations. It is changing the face of Europe in spite of the prevailing frontiers, migration laws, racism and xenophobia. The question that remains to be answered is that of how long whiteness can endure after the fall of white primacy.

I consider it important to shift the gaze from the source to the structure within historical analyses. This requires intentional action to force the con-text, namely the device through which history as well as every other discipline within the humanities is created, in order to produce new meanings and interpretations. "To force" does not mean to encapsulate stories within the history; this is a method applied to understand how borders operate through subjectivation and consequently through power in the reproduction of knowledge.

I consider this approach important for demonstrating a correspondence and interaction among symbolic, physical and narrative borders in Europe. My aim is to focus on practices of history making that can problematize and criticize the narrative order and its role in reproducing the discipline. This constitutes *border work*, the aim of which is neither to write another history of the Black Mediterranean nor to reproduce the European logos. This brings me to a discussion of some of these practices and their relevance in problematizing the matter at hand.

One of the most powerful practices entailed in the making of history concerns the act of writing. There is a specific style of writing history impersonally that while not formalized is so prevalent as to be canonized. This means that first, the historian, namely whoever writes history, has a specific positionality and never appears in the text. Second, an "unstated fact" is evident, according to which "I" and "me" shall not appear in the text, thereby constituting what can be termed a tradition that enables the continuation of the work of previous scholars. This approach situates history in the past and questions the knowledge produced as if it is, mandatorily, of importance for interpreting the present. This ontological scheme affirms and confirms the centrality of the colonial archive, including when the narrative entails an attempt to denounce stereotypes and categories.

Therefore, I think it is essential to declare the positionality from which the historian speaks. Thus, in my case, I am a white scholar at a European university, without a tenure track position, and I work for a Portuguese foundation. This means, among other things, that attention should be paid to what is happening outside of the context, shifting the gaze from the centre to the periphery of the topic. In particular, it is important to focus attention on the scholar's subjectivity and its transformation during the process of fieldwork. The deconstruction of subjectivity and the expected research outcomes could simultaneously constitute a process of decolonizing history. This happened to me in Ventimiglia, which is located on the Italian-French border. Upon my arrival at the frontier, my plan was simply to collect oral interviews with "migrants" and Italian or French citizens. However, with the passage of time, I realized that what was happening around me was part of a global history that was individually lived by thousands of people on a daily basis. At a certain point, I understood that I too was part of that history and that the border was, first of all, within me (in terms of my ideas and conception of the world).

Following this passage, I sought to decenter the context, to examine its border. The question that I focused on is what remains outside of the investigation of history. I immediately contemplated a history of silences generated by those people who daily cross a border *within* or *towards* Europe. In order to analyse this aspect of the Black Mediterranean, I applied oral history and the tool of intersubjectivity, with the understanding that an individual meaning can be assigned to each silence, or, on the contrary, a common or similar value can be assigned when it is analysed or appears in the course of an interview. The key role of empathy, which is a language capable of transcending the order of language, lies in its contribution to enabling an understanding of the role of silences within a story (Proglio 2018a, b, c).

I applied the same approach to investigate migrant personal itineraries in Europe. At the end of each interview, I asked the interviewee to draw a map of his/her trajectory from the place of departure to Europe. This tool was used during the "Bodies Across Borders: Visual and Oral Memory in Europe and Beyond" research project led by Professor Luisa Passerini, funded by the European Research Council. Within that framework, I was working on a specific project dedicated to constructing an oral history of the diaspora of the Horn of Africa within Europe through interviews conducted with individuals who had arrived in Italy via different routes, with some having crossed the Mediterranean. Deconstructing these drawings in combination with the interview transcripts, I concluded that the condition of being in diaspora could produce multiple temporalities and, consequently, a huge transformation in the conception of the past, and in the approach to history making (Proglio 2019).

From a perspective in which history is considered a cultural process that is concerned with the role of a nation, I have tried to focus on another aspect that relates to public memory: the legacy of colonialism in the Mediterranean. The key question guiding my inquiry is as follows: Is colonialism only concerned with the history of European nations outside of their frontiers? This may appear to be a silly question, but it is not. If a native is reduced to his or her representation, and is dehumanized and individualized as Fanon has contended and foreclosed according to Spivak, the question of what the impact of this knowledge production is on the postcolonial condition assumes particular salience. Many scholars engaged in postcolonial studies have argued that the colonial past has been repressed because of the loss of colonial dominions. But, if these memories have in fact been repressed, how could xenophobic and racist stereotypes reappear in the present in connection with "migrants?" Does a declaration that mainstream history, namely one centred on the roots of the national community, is a history dominated by racism, violence and power relations forged along the line of colour signal the existence of shame within the national public imaginary and individual consciousness? From this perspective, I consider an investigation of the term "repression" in which multiple gazes on pasts and presents are analysed to be highly strategic and aimed at decolonizing this lack of memory. Stereotypes that were framed during the colonial era appear, surprisingly, each day on the shores of Europe, marked on other black bodies (Proglio 2018a, b, c).

Conclusions

- We... blacks . . . we . . . Africans . . . do not have a history—not yet—in Europe. We lost all on the other shore of the Mediterranean. Europe stole all from us, our bodies, our pasts . . .
- Why are you saying this, Rachid?
- Because my name, and all of our names, have no meaning
- [*Silence*]
- Our silences and voices are on your lips and [in your] hands

This extract from a dialogue with a man in Ventimiglia, on the Italo-French border, reverberated strongly for me, in my soul, after the fieldwork was concluded. Europe is at war against these people termed "migrants." Laws on citizenship, walls, frontiers and xenophobic attacks are all tools deployed in this war, which is based on a violence whose roots can be found in the colonial past.

This journey across the Black Mediterranean, conducted through the thoughts of Fanon, has illuminated many similitudes and connections existing between the condition on the "native" of yesterday, in the colony, and the postcolonial contemporary condition of the "migrant" in Europe. In particular, it has revealed the importance of conducting an analysis of the role of violence to elucidate, in depth, how Europe acts, using which tools, to confirm and reiterate its centrality. One of these tools, among others, is the line of colour and its proliferation both within and beyond European frontiers. This border has been embodied and imposed on the skins of thousands of people.

In my research, commencing with a declaration of my own positionality, I have attempted to problematize how borders work in the construction of history; a history that as Fanon pointed out, centres on the role of Europe and those of its nation states, languages and narratives. In this endeavour, my first goal was to present a critique, aimed at simultaneously revealing the existence, outside of the discursive order and beyond representations of the native/migrant.

Given the reverberation of colonialism with what is currently happening in the Mediterranean, I would argue that the use of knowledge to deconstruct, decolonize, and provincialize the canonized narrative on history is essential. That action will not render Rachid and others like him visible, for its aim is not to give them a voice. Rather, the intention is to take up a position in a war in which violence has been normalized by European institutions and to attempt to elaborate a history without borders wherein subjects and subjectivity precede the context.

References

Anderson, B. 1983. *Imagined Communities*. London: Verso.

Arendt, H. 1951. *The Origins of Totalitarianism*. London: Schocken Books.

———. 1958. *The Human Condition*. Chicago/London: The University of Chicago Press.

Beneduce, R. 2004. *Frontiere dell'identità e della memoria. Etnopsichiatria e migrazioni in un mondo creolo*. Bologna: Franco Angeli.

———. 2007. *Etnopsichiatria: sofferenza mentale e alterità fra storia, dominio e cultura*. Roma: Carocci.

———. 2013. Un corpo che interpella, o la cura della storia. In *Fanon postcoloniale. I dannati della terra oggi*, ed. Miguel Mellino, 31–49. Verona: Ombre corte.

———. 2014. *Archeologie del trauma: un'antropologia del sottosuolo*. Roma: Laterza.

Chambers, I. 2013. Il ritmo del tempo: sempre Fanon. In *Fanon postcoloniale. I dannati della terra oggi*, ed. Miguel Mellino, 11–19. Verona: Ombre corte.

———. 2019. *La questione mediterranea*. Milano: Mondadori.

Cuttitta, P. 2012. *Lo spettacolo del confine. Lampedusa tra produzione e messa in scena della frontiera*. Milano: Mimesis.

Danewid, I. 2017. White innocence in the Black Mediterranean: hospitality and the erasure of history, Third World Quarterly, 38 (7): 1674–1689.

De Genova, N. 2013. Spectacles of Migrant 'Illegality': The Scene of Exclusion, the Obscene of Inclusion. *Ethnic and Racial Studies* 36 (7): 1180–1198.

Derrida, J. 1991. Des Tours de Babel. *Semeia* 52: 3–34.

Du Bois, W.E.B. 1903. *The Soul of Black Folk*. New York: New American Library.

Fanon, F. 1952. *Peau noire, masques blancs*. Paris: Editions du Seuil.

———. 1959. *L'An Cinq de la Révolution Algérienne*. Paris: Maspero.

———. 1961. *Les damnés de la terre*. Paris: Maspero.

———. 1964. Racism and Culture. In Id. *Toward the African Revolution. Political Essays*, 29–45. New York: Grove Press; Translated from the French by Haakon Chevalier.

Ficek, D. 2011. Reflection on Fanon and Petrification. In *Living Fanon, Global Perspectives*, ed. Nigel Gibson. New York: Palgrave Macmillan.
Foucault, M. 1971. *L'ordre du discours*. Paris: Gallimard.
———. 1977/1980. *Confession of the Flesh, in Power/Knowledge: Selected Interviews & Other Writings* 1972–1977, ed. Colin Gordon, 194–228. New York: Pantheon Books.
———. 1998. *The Will to Knowledge: The History of Sexuality*. Vol. 1. London: Allen Lane; Original Edition Gallimard 1976.
Gilroy, P. 2004. *After Empire: Melancholia or Convivial Culture*. London: Routledge.
Giubilaro, C. 2017. Lo spettacolo del naufragio. Migrazioni, luoghi visuali e politica delle emozioni. In *Visualità e (anti)razzismo*, ed. Elisa A.G. Afrini, Valeria Deplano, Annalisa Frisina, Gaia Giuliani, Mackda Ghebremariam Tesfaù, Vincenza Perilli, Alessandro Pes, Tatiana Petrovich Njegosh, Gabriele Proglio, and Daniele Salerno e Alessio Surian, 10–24. Padova: Padova University Press.
Gramsci, A. 1971. *Selections from the Prison Notebooks of Antonio Gramsci*. London: Quintin Hoare and G. Smith.
Guha, R., ed. 1982. *Subaltern Studies 1: Writings of South Asian History and Society*, 7 Vols. Delhi: Oxford University Press.
Hall, S. 1980. Race, Articulation and Societies Structured in Dominance. In *Sociological Theories: Race and Colonialism*, ed. UNESCO, 305–345. Paris: UNESCO.
———. 1997. *Race, the Floating Signifier*. Northampton: Media Education Foundation.
Loomba, A. 1998. *Colonialism/Postcolonialism*. London: Routledge.
Mazzarra, F. 2019. *Reframing Migration. Lampedusa, Border Spectacle and the Aesthetics of Subversion*. Oxford: Peter Lang.
Mbembe, A. 2003. Necropolitics. *Public Culture* 1 (15): 11–40.
Memmi, A. 1967. *The Colonizer and the Colonized*. Boston: Beacon Presso.
Mezzadra, S. 2008. *La questione postcoloniale. Storia e politica nel presente globale*. Verona: Ombre corte.
Proglio, G. 2018a. Is the Mediterranean a White Italian-European Sea? The Multiplication of Borders in the Production of Historical Subjectivity. *Interventions* 20 (3): 406–427.
———. 2018b. Silences and Voices of Mediterranean Crossings: (Inter)Subjectivity and Empathy as Research Practice. *Revista Brasileira de Pesquisa (Auto) Biografica* 03 (7): 67–79.
———. 2018c. L'Italia e il passato colonial. Riflessioni e considerazioni a margine del dibattito storiografico. *Memoria e ricercar* 1: 113–132.
———. 2019. Conceptualizing Diasporic Memory: Temporalities and the Geography of Emotions in Eritreans' Oral Tales. In *The Mobility of Memory*

Across European Borders. Migrations and Diasporas in Europe and Beyond, ed. Luisa Passerini, Milica Trakilovic, and Gabriele Proglio. London: Berghahn.

Pugliese, J. 2010. *Biometrics: Bodies, Technologies, Biopolitics*. New York: Routledge.

Said, E. 1979. *Orientalism*. New York: Vintage Books.

———. 1993. *Culture and Imperialism*. New York: Vintage Books.

Sayad, A. 1999. *La double absence*. Paris: Èditions du Seuil.

Siebert, R. 2012. *Voci e silenzi postcoloniali. Franzt Fanon, Assia Djebar e noi*. Roma: Carocci.

Spivak, G.C. 1988. Can the Subaltern Speak? In *Marxism and the Interpretation of Culture*, ed. C. Nelson and L. Grossber, 271–315. Urbana: University of Illinois Press.

———. 1999. *A Critique of Postcolonial Reason: Toward a History of the Vanishing Present*. Cambridge: Harvard University Press.

Stoler, A. 2008. *Along the Archival Grain. Epistemic Anxieties and Colonial Common Sense*. Princeton: Princeton University Press.

———. 2016. *Duress. Imperial Durabilities in Our Times*. Durham: Duke University Press.

Todorov, T. 1982. *The Conquest of America: The Question of the Other*. New York: Harper & Row.

Wekker, G. 2016. *White Innocence: Paradoxes of Colonialism and Race*. Durham: Duke University Press.

CHAPTER 5

Colonial Cultural Heritage and Embodied Representations

Giulia Grechi

Throwing the Body into the Struggle[1]

That's why I just want to live
Even while being a poet
Because life expresses itself by just being.
Examples will be my form of expression.
Throwing my body into the struggle (…)
As a poet, I will be a poet of things.
Life in action will be my expression,
Poetry will be life in action,
Because, again, there is no poetry without real action. (Pier Paolo Pasolini, *Poeta delle Ceneri*, 1966)

In Sweden since the end of the 1990s, hundreds of children of asylum seekers from the former USSR, the Balkans, and the Middle East, children who have lived in Sweden for months or even years waiting for their

[1] Thanks to Georgia Wall for the translation of this chapter.

G. Grechi (✉)
Cultural Anthropology, Academy of Fine Arts Brera, Milan, Italy

G. Proglio et al. (eds.), *The Black Mediterranean*, Mediterranean Perspectives, https://doi.org/10.1007/978-3-030-51391-7_5

residence visas upon being denied political asylum, fall seriously ill. They go into a coma that can last months or years, literally losing the will to live. They just fall away from the world.

This illness has been dubbed ‘Resignation Syndrome’. Since 2005, more than four hundred children between the ages of eight and fifteen have been affected by this mysterious condition, some of whom have not recovered even after having been granted the right to asylum. These are not young people with previous neurological or mental health problems, nor are they feigning illness. Rachel Aviv, in a report for The New Yorker, tells the story of Georgi, a Russian refugee who arrived in Sweden with his family when he was five years old (Aviv, Rachel. 2017. *The Trauma of Facing Deportation*. The New Yorker. https://www.newyorker.com/magazine/2017/04/03/the-trauma-of-facing-deportation. Accessed on 14/09/2019). In 2015 the Migration Board rejected his family’s claim for political asylum, and they were to be deported to Russia within months. Georgi fell into a coma for seven months. In an open letter to the minister of migration, a group of Swedish psychiatrists denounced the proliferation of similar cases and accused the government of ‘systematic public child abuse’. They suggested that Resignation Syndrome was provoked by two overlapping traumas: first, the violence the children had suffered in their homeland, and second, the terror of having to return, often with their lives at risk. The Swedish Board of Health and Welfare suggested that “the most effective ‘treatment’” (ibid.) of Resignation Syndrome is to grant permanent residency. Dr. Elisabeth Hultcrantz, who cares for many of these patients, interprets their symptoms as part of a mental health disorder caused by the sudden perception of a lack of security and lack of hope for the future. Speaking specifically about Georgi, she declared, “If the boy does not have security, he will not wake up in whatever country he is in” (ibid.). Georgi’s family, as a direct consequence of his illness, was granted permanent resident status in 2016 after seven months. After a few weeks, Georgi began to recover.

These cases raise several questions. I suggest here that we need to find a way to read these symptoms as symbols, narratives, expressions of the rejection of a condition of subordination to which these subjects seem to be condemned and against which they seem to have no way to fight back. We must understand the ‘infrapolitical’ dimension to recognize the mix of political positioning and psychic suffering that these bodies express through their symptoms.

Not recorded in any other country
not recognised as an illness, not a true
coma

Georgi's case is not isolated. Djeneta and Ibadeta, two Roma sisters from Kosovo, were both in comas at the time of writing. Djeneta is twenty years old and has been in a coma for two and a half years; Ibadeta, fifteen, in a coma for five months, both as a result of being denied asylum by the Swedish government. They were violently attacked in Kosovo because they are Roma, and so fled to Sweden. When they arrived in Sweden they had no evidence, no *document* attesting the violence they had suffered, and so their appeal for political asylum was rejected. But, crucially, Djeneta and Ibadeta *are* the document: these girls, perhaps, are *embodying* the fear, threat and violence they associate with having to return to Kosovo, and their coma is a way to communicate the unspeakable but essential circumstance of their lives: "they feel totally helpless, and they become totally helpless" (ibid.).

One pressing question is whether there is a way to cure these people, these "body-selves," (Scheper-Hughes 1994) these bodies that express subjectivity and will through symptoms, that do not give up but, depriving themselves of humanity, enact a form of extreme resistance via the coma. But there is an equally pressing question which regards us more directly: Who and what should we cure?

Of course, these young people need to be cured—but European migration and reception policies also need to be fixed. Perhaps the real 'disease' is 'fortress Europe', which has transformed the basic right of all human beings to move freely into the privilege of a few. Perhaps this unprecedented form of cultural violence which preaches rejection of others and which forms the basis of European migration policies should be stopped, this intolerance of the Other that has spread throughout Europe and beyond.

I use these stories as a departure point. I want us to shift away from the dominant point of view (of our societies, institutions, and daily practices), which understands corporeity in relation to power. We know a lot about how power exercises its own discourse on bodies through technology and biopolitics, prohibition and pleasure. We accept how each culture constructs its own meanings, its own ways of thinking about the body, the normative forms and models of each society, so that people belonging to that culture or society can feel 'normal' by corresponding to *the norm*, in contrast to those 'Others' who do not. We are familiar with the "body techniques" Marcel Mauss (1965) describes, which we absorb from when we arrive in the world because they are the practices through which our bodies think and act in an acceptable way within our social spaces, the practices which make our body much more 'cultural' than 'natural'. Of course, we eat, sleep, breathe, and go to the bathroom because it is 'natural'—we would die otherwise. The point is *how* we do it: in which rooms, positions, with which gestures and rituals and taboos, individually or collectively, and what meaning we give to it all.

We know how our bodies have incorporated and continuously re-enact the script of power as it is conceived in our culture: through certain postures, a certain social distance, ways of touching or looking. We know a lot about this. Our ethical and political responsibility here is to reverse the question: rather than asking what power knows about the body, we should ask what the body knows about power. We must ask ourselves this question because it can help us listen to the bodies of Georgi, Djaneta and Ibadeta, to understand the language they use to express their reason, and to deeply interrogate our own.

I would therefore like to weave together various discourses, including the relationship of the body with power, memory, and urgent contemporary matters concerning those perceived as 'the bodies of others', or simply, 'others'. I refer specifically to the cultural legacies of Italian colonialism and our current difficulty in contemplating the Other as a subject, in the humanity that unites us. Let us begin with Achille Mbembe's challenge:

> The theoretical and practical recognition of the body and flesh of "the stranger" as flesh and body just like mine, the idea of a common human nature, a humanity shared with others, long posed, and still poses, a problem for Western consciousness. (Mbembe 2005, p. 8)

To weave together this web, we must have the patience step inside the mirror. In this chapter, we will find ourselves confronted with two very different artistic works: two thresholds, two different ways that affect us, two gazes which interrogate us.

(Post)colonial Remains

> *The past that does not want to pass, evoked by many contemporary historians, is a past which lives on in the present, or rather, haunts it, an ever-present ghost.* (Ricoeur 2004, p. 83)

We are indeed surrounded by the spectral traces of our colonial history. They are everywhere: in plaques and monuments, in the narratives of museums (particularly in ethnographic museums, which are still struggling to come to terms with their original sin of complicity in colonialism), in the architecture and toponyms of entire neighbourhoods. In Rome, the city where I live, we have the 'African quarter', with its Viale Etiopia, Piazza Amba Alagi, Via Tripoli, Via Asmara—an inscription of space that echoes and resonates with that *colonial elsewhere* that has structured our cultural imagination of otherness and ourselves, a whole colonial order of discourse that orients and forms a background to our daily lives.

Even our houses, our mirrors, are haunted by ghosts—or rather, by spectres. A ghost is an illusory image, an imaginary projection. Spectres instead have a material body: audible, touchable. They may be partially invisible, but never incomprehensible. In etymological terms, *spectrum* means both sight and the tool for seeing. A sensitive image is at the same time the way to see something else. Spectres are therefore alive, and they speak. They are full of signs. Agamben, in his short essay *Dell'utilità e degli inconvenienti del vivere fra spettri* (On the Uses and Disadvantages of Living Among Spectres, 2009), says that spectres are composed of signs,

> Or rather, more precisely, of signatures, that is of those signs, figures or monograms that time scratches on things. A spectre constantly carries a date in its wanderings and is therefore an intimately historical being. (Agamben 2009, p. 61)

It seems that, in relation to their colonial histories, most European countries have developed the "memory diseases" that Paul Ricoeur (2004) talks about: they remember too little, and not well, in a sort of active forgetfulness, an oblivion of escape. It is a wilful not-remembering, or a desire to remember in a certain way. This "selective oblivion," as Ricoeur calls it, has governed memorialization strategies in many countries, structuring what and how to remember: strategies that must now be critically questioned. In Italy, the myth of Italians as *brava gente* (Del Boca 2005) and its 'minor' colonialism compared to that of other European countries were constructed over the thirty years between the fall of Fascism (and with it Italy's colonial venture) and the release of documents relating to colonial campaigns in Ethiopia, Eritrea, Libya, Somalia and Albania. In a strategic, criminal delay, these archival documents only became available in the 1980s. During those long years of silence, Italians are said to have 'forgotten' their colonial history. I believe that it has not been not so much forgotten or erased, but rather repressed.

Continuing to think in psychoanalytic terms, repression occurs when the traumatic event is not erased from consciousness, but moved to another 'place', to another affective constellation. This cancels the violent and racist root of colonialism and moves this complex event into the realm of the exotic, the paternalistic, the 'place in the sun', the 'African dream'. Amnesia, active or selective forgetting, the construction of mythological imagination, and repression itself can all be considered provisional strategies to confront a difficult memory. This memory may be tied up in the conflictual relationship between *History*—the institutional narrative of those events—and *stories*—the intimate memories that many of us have of that past (and that often cannot be reconciled with History). All this generates certain imaginaries about ourselves and others, giving shape to those racist and violent drives, directed especially towards immigrants' impulses that remain inexplicable and incomprehensible if we cannot connect the traces of those movements associated with one place, with another memory. But what is the 'place' of this memory?

The colonial traces I mentioned previously are not only found in public spaces and in the institutional narratives of our cities. They are also in the privacy of our homes, in drawers or attics, in old boxes of photographs of a grandfather who went to war in Africa or in letters from those who emigrated to the colonies, or in our cupboards: in Tripolini biscuits, Assabesi liquorice, Negrita chocolates, Carta d'Eritrea perfume sachets; in 1960s toys like *Ciccio Bello Angelo Negro* or *Bessie the lively little negro*; in hit records like "Bongo" from 1948, or "I Watussi" from 1963; or the nursery rhyme about the Black man, "*la Ninna nanna dell'uomo nero*," sung to many of us as children. Colonial traces are in the sugar sachets their racist iconography offered in coffee bars and in the racialized imaginaries that still circulate among advertisements, music videos, fashion photography and everyday languages, such as the expression "*sono incazzato nero,*" a way of saying "I'm furious," but literally meaning "I'm black angry."

In the terms of Vincent Meessen, a Belgian artist and curator, this is a real "colonial hauntology," (Gregos and Meessen 2015, p. 10) an uncanny process of the appearance of colonial discourse and its imaginaries in unexpected places, in the daily intimacy of gestures, words, and objects. Popular, racialized culture of the colonial period is very much alive and active, albeit in disguised or seemingly harmless forms and thus difficult to unmask. This creates a "dizzying proximity," as Merleau-Ponty (2002, p. 39) puts it, between things, bodies and memories, making the process of remembering itself an essentially political act, a claiming of space and body, a re-member-ing, (in Italian, *rimembrare*), a recomposition of the severed limbs of our cultural identity (Grechi and Gravano 2016, p. 42).

Colonial imaginaries and representations are like an underground river, invisible but ubiquitous. They are within our mouths, our digestive systems, our ears and our voices. We have inhaled them in certain smells, and they have nourished us and imbued our white skin. They are the gloves with which we touch, the *coloured* lenses that filter our reality. We have, quite literally, *incorporated* them. They have become *habitus*, habits, clichés, stereotypes, parts of ourselves and our identity. The body is the main place of this memory, a memory which is translated and reproduces itself through us. Paul Connerton reasons that "in habitual memory the past is sedimented in the body" (Connerton 1999, p. 87). Memory (individual and collective) is not only spiritual, intellectual or semiotic recall within a certain cultural tradition, but has to do with the body. To understand how the memory of a social group is constructed, transmitted and nurtured, it is necessary to connect the process of remembering with the bodies that do it. In this way, certain forms of knowledge are reproduced through a series of rituals, attitudes, gestures and habits, or "affective predispositions" (ibid.), and through the performativity of the bodies which stage them. Memory, even collective memory, is always embodied. And it is always cultural, just as the bodies that animate it are cultural beings. Each one of us has absorbed a sort of choreography of power and authority, expressed through gesture and behaviour and inscribed with a specific cultural meaning—gestures and behaviour that are not consciously taught or communicated, but simply reproduced to a point that they become so automatic and 'infra-ordinary' (Perec 1994) that we take them for granted. They seem transparent, innocuous, and natural. In this sense, Connerton explains, the body is a real-life mnemonic system. Its performativity automatically recalls an entire "discourse order" related to power, identity, and difference. Thus, "every social group attributes specific values and meanings that it hopes to preserve to physical reflexes to corporeal automatisms" (Connerton 1999, p. 117). The body understood in this way is not a Cartesian *rex extensa*—an object, a container of mental concepts—but a subject producer of meaning, a *body-self*. The body is mindful (Scheper-Hughes 1994). All of our knowledge, including memory and its imaginaries, is embodied, produced by and through the body. The question we need to ask ourselves at this point is:

> The body is the overdetermined point where differences collide, the epidermal surface on which racism etches its mark, and a ground of resistance from which alternative counter-narratives can be produced (…). In the ritual exchange of stereotypes of the body between 'race', gender and sexuality, racism and cultural differentialism had deployed their most violent and destructive fantasies. How could this be contested and undone? (Hall and Sealy 2001)

If the body *does* all this, the body *knows* all this. That is to say, it is within our body-self that we can find the key to deconstruct, contest and subvert the incessant repetition of these imaginaries. Now we are inside the mirror. It is precisely here, in this tangle of memories and embodied representations of contradictory affects, that an artistic approach can be meaningful. It can bring out, give body and visibility to those spectres that remain invisible, allowing us to go back and explore our colonial history and the roots of our identity, our colonial unconscious. The critical space opened up by artistic and poetic language can invite us to question ourselves, striking at the heart of our discomfort and our intellectual security, our *common* sense, our complex affects.

Mum, I'm Sorry

> *Go away, away, but not by land*
> *Sail away, sail away, but not by sea*
> *Fly away, fly away, my dear,*
> *but do not touch the air.* (Wislawa Szymborska, *Discorso all'ufficio oggetti smarriti*, 2004)

The film *Mum, I'm sorry*, by Martina Melilli,[2] was born out of a dialogue with migrants who survived the long journey over land and sea that brought them to Italy (for now, at least). It was also as a response to the work of Dr. Cristina Cattaneo, a pathologist and forensic anthropologist whose task has been to identify the bodies of those who die at sea or before getting to the Sicilian coast. In her role as a coroner, she tries to reconstruct the stories and identities of people from the signs on their corpses, committed to a humanitarian forensic medicine. From these points of departure—the bodies of those who do not survive, from their

[2] www.martinamelilli.com

stories and lost, ignored identities—Martina Melilli ponders how to tell what seems untellable, yet what *must* be told.

Surprisingly, despite the forensic focus on the dead, no bodies appear in Melilli's film. On the autopsy table, only objects are examined—but not just any objects. They are objects found at sea after two shipwrecks off the coast of Lampedusa in October 2013 and May 2015. They are the migrants' most important objects, those chosen from many for a difficult and dangerous journey. They are objects of the greatest value and significance, the material shoring of an individual's intimate identity, signs of memories and affect, or tools that can facilitate life upon arrival (such as telephone numbers, addresses, references, and qualifications). Shipwrecked objects, collected and sealed in plastic bags, they have lost their use-values and meanings but not their ability to tell. In the film, it is particularly striking that we see these things through a clinical/investigative/bureaucratic frame: as traces and evidence through which we try to identify, reconstruct and catalogue.

I have an ambiguous reaction to the anonymous hands touching, moving, investigating these "bodies of evidence" in Melilli's film. In a sense, I am frustrated by the cold, detached nature of the operation: by those latex-covered fingers which seem to violate the intimacy of unknown people; by the invisibility of the subject who has the power and authority to investigate, handle, catalogue, define (an increasingly impersonal, inhuman power); by the violence of the procedures, the operations and the regulations that organize and govern migration through an insistent dehumanization of people, translating into ruthless daily practices "that violence of History that you are often invited to accept with resignation, as an unavoidable reality" (Beneduce 2010, p. 4). But I also feel that those hands are looking for something: they are trying to reconstruct the stories locked in those objects. The only way possible is to examine the traces which continue speak of the person to whom they belong. The archaeological gaze fixes a microscopic attention on every detail of the object, excavating, almost, its essence. The salt-encrusted photographs reveal shreds of everyday life: memories of parties, houses, families, and friends. Clocks that still tell the time, contraceptive pills, documents, address books, letters. Like the bodies that Dr. Cattaneo tries to identify on her autopsy table, these objects are *places of identity*, object-bodies, brimming with traces and revelations. They are places of a memory incorporating a complex mesh of intimate violence, structural violence (public, historical, political) and above all, *desire*. In *Black Skin, White Masks*, regarding the experience of the colonized, Frantz Fanon wrote, "as soon as I desire I ask to be considered. I am not merely here and now, sealed into thingness. I am for somewhere else and for something else" (Fanon 1996, p. 190).

Discourse on migration is strongly rhetorical. At worst, it creates a rhetoric that associates migration with criminality; at best, with a sense of loss of leaving one's native land, losing one's identity, renegotiating one's own cultural traditions, abandoning one's personal affects. But whoever embarks on such a perilous journey is not only fleeing from something. They are also driven by desire and aspiration. According to Arjun Appadurai (2011), it is not only economic resources that are unevenly distributed throughout the world, but also the capacity to aspire. This ability …

> … like any other complex cultural capacity, survives and flourishes only if it can be practiced, repeatedly used and explored by the elaboration of hypotheses or disputes: where the opportunities to formulate hypotheses and contestations regarding the future are limited […] it follows that the capacity to aspire is relatively less developed. (Appadurai 2011, p. 23)

The possibility to desire and build aspiration becomes for many a true need.[3]

It seems we have assimilated injustice: we are no longer able to identify and fight it, or even recognize with humanity the experience of others. We are victims of what Didier Fassin calls *cultural anaesthesia*, which makes "the suffering of the Other in public culture and discourse inaudible […], renders us insensitive to the fate of others and most significantly makes them seem incomprehensible" (Fassin 2016, p. 12). The deluge of media images of dead, wounded, hungry, sick, and homeless bodies makes us on the one hand aware of global inequality, but on the other, it "constitutes a form of depersonalisation of others that renders them simultaneously neighbours and strangers."[4] In Martina Melilli's film, silence is broken only by the constant annoying background buzz, perhaps an oblique reference to how migration discourse is often perceived. From the cold 'voice' of the technology with which we 'map' the Other to the mechanical and muffled submerged sound of boat engines, which likely sounds as

[3] This insight emerged during a discussion with Roberto Bertolino, a pyschotherapist at the Centro Fanon in Turin, as part of the seminar *Etnopsichiatria in azione. Ripensare l'altro nei luoghi dell'accoglienza e della cura*, Centro Fanon, Torino, 17–18 marzo 2018.

[4] Ibid.

they do to those in the cargo hold, the sound in this video invites the viewer to identify with different positionings and different points of view.

Where, then, lies the place (and responsibility) of the viewer, of all of us spectators, not only of this film, but of history and its violence? Though the atmosphere of the film is cold and clinical, the grammar of the film does not allow the viewer to step away and remain detached. The hands that touch those objects ultimately represent our own, and we are implicitly called to decide whether we identify with the subject of power, who catalogues, analyses and archives; with the subject, who cares and seeks to reconstruct and to recognize; or with the Other, who is presented in a fragmented, spectral form, through traces and absences. The viewer cannot remain at a 'safe' distance.

The film 'touches us', in a sense. It puts us in touch with how we think about ourselves in relation to the Other, to those who we perceive as strangers. It provokes that uncanny sensation of recognizing familiarity in those objects that belong to people we think aren't related to our lives. Those objects could belong to any of us. They are the same things that crowd and comfort our daily lives, and in this way, we can recognize humanity in estrangement, the self in the Other. On an emotional level more than a rational one, this film urges us to recognize the representations and memories that we have incorporated and that have shaped our knowledge about ourselves and the world.

When the camera zooms out, away from the close-ups of bodies-of-things, we see silent, nocturnal landscapes of riverbanks and bushlands. These are places of arrival or escape, barely illuminated by a single light which pans rhythmically across the scene, like the searchlight of the coastguard. Then, one of these landscapes is revealed as a Black body. Up close, we regard skin, hands, hair. Object-bodies and body-landscapes until the camera reaches an open eye, staring at us through the camera, returning our gaze: an *eye/I*, fixedly interrogating us, telling us how much this does regard us.

As Far as My Fingertips Take Me

May your body always be
a beloved space for revelations. (Alejandra Pizarnik, *Los Trabajos y las Noches*, 1965)

We go into the room one at a time. As soon as I enter, I am asked to put on a white coat and to sit in a chair facing the window. To my left, there is a white wall with a hole at shoulder height. I put on the headphones I have been given and listen. A male voice begins, "Hello. My name is Basel. I'm the other side of the wall." Basel asks me to put my arm through the hole. I can't see what is on the other side; the hole is just wide enough for my arm to fit in. Basel starts to tell me about himself and his journey: he is a political refugee. Simultaneously, Basel's hands gently take my fingerprints, one by one. Then he starts to write or draw something on my arm. His touch is light, warm, sensitive. Basel continues telling his story, and he asks me to listen to a song: a rap inspired by the journey his sisters made from Damascus to Sweden. His hands continue to hold my arm, to delicately trace something I cannot see. Then, he tells me he's finished and that I can withdraw my arm and leave the room. His hands guide my arm. A moment before I extract my arm, my hand cannot help but grip his, to hold it and keep us for a moment longer in such intimacy. When my arm is reunited with my eyes, finally I can see. My fingertips are marked with ink. From one of them, a thin thread keeps a small boat adrift in the palm of my hand. A long line of small figures walks up my arm; their destination is a mystery, even though it runs across me. Adults and children, some with a backpack slung over their shoulders. Like a landscape in miniature seen from a distance: they are silhouettes of people, almost shadows. I cannot glean any more, only the evidence of their steps. Their path seems to stop in the crook of my elbow, where they find themselves confronted by a peremptory line, a wall that interrupts their journey and cuts into my veins at a right-angle. A wall separates me from Basel in that room. Many walls, some material, other less so, interrupt the travels of many people in many parts of the world, and one of them is now on my skin.

History is always written on bodies. And sometimes our body hides behind its intact skin. *As Far As My Fingertips Take Me,* the performance I described above by Tania El Khouri,[5] is a performance which is partially invisible, but certainly not incomprehensible: "Through touch and sound, this intimate encounter explores empathy and whether we need to literally 'feel' a refugee in order to understand the effect of border discrimination

[5] I participated in Tania El Khouri's performance during the Festival of Santarcangelo di Romagna, 7–14 July 2018.

on peoples' lives" (http://taniaelkhoury.com/as-far-as-my-fingertips-take-me/. Accessed on 14/09/2019).

The skin is our largest organ, separating and protecting us from external agents, but simultaneously allowing us to perceive and touch. Our fingertips are a hub of perception: they allow us to touch and feel, to understand beyond sight, through tactility. Yet they are also a site of biopolitics, of identification and control, a unit of measurement of a subject's identity and of the possibilities or limits that follow. It is via a large fingerprint database that different European countries exchange data on migratory flows. Fingertips are the main subject of a European law, the Dublin Regulation, which means a refugee's journey can be set as far as their fingertips take them. Fingerprints are units of measurements of a subject and, if born beyond the borders of Europe, they delineate the space where that subject can move. They can be a condemning sentence, therefore, because this law entails that an immigrant or refugee stopped anywhere in Europe can be sent back to wherever their fingerprints were first recorded, regardless of their plans, aspirations, or the people they were trying to reach.

Tania El Khoury's performance overturns the public and media discourse on migration that we have become used to by bringing it back to a human dimension, linked to the experience of individual living beings, each with their own history and suitcase of dreams. It is in the moment when we literally 'feel' the experience of a single story like Basel's that we are able to retrieve an empathy that goes beyond a simple identification, because it passes through a process of profound and radical embodiment. Thus, we are able to understand emotionally and access a knowledge of what it is to be human through what Renato Rosaldo calls "the cultural force" of emotions (Rosaldo 1989). It is through this complex bodily feeling, packed with contradictory emotions, that we are able to experience the presence of 's body. George Devereux's reflections on the 'body-to-body' between researcher and researched seem telling here, applicable as they are to any meeting with someone perceived as an 'Other': the distance we project onto this encounter (the sense of detachment and alleged objectivity this should guarantee) inhibits "even the creative conscience of solidarity" (Devereux 1984, p. 277) between ourselves and this other, a defensive refusal to recognize a common human nature and a shared humanity.

Perceiving the body of the Other, its singular history as a human and embodied experience (rather than in the generalized disembodied experience of 'migrants' as an anonymous and stereotyped collective), instantly eliminates any sense of distance. After I left the room, I tried to capture the rush of sensations I experienced in a note: "it was an unexpected contact, 'just' a handshake, it was a feeling of being (able to be) together." Tania El Khoury's performance insists on the 'infraordinary' of the intimate relationship between two human beings, conscious that this is the site where we can learn to recognize our vulnerability together with the ethical and political dimension of our daily thinking-talking-acting-living. Misrecognition of the Other leads to racism: "it is not an individual disease, but a power structure that continues to generate the hierarchization of the world" (Chambers 2018, p. 7).

It is not the aim of contemporary artistic works like Tania El Khoury's performance and Martina Melilli's film to represent—this is the task of anthropologists, sociologists and historians. Rather, they can help us notice an impossibility, a contradiction, the lapse which reveals a truth we are afraid to recognize because it makes us vulnerable and forces us to question our positions of privilege. These works help us to experience an opacity, particularly in terms of rejecting the spectacularization of the drowning, delinquent, disobedient migrant body, the body that rejects being categorized and tracked and thrusts the gap between law and justice in our faces. The opacity recorded by these works from one side is enforced, as in the case of the shipwrecked bodies, which remain absent and nameless, present only as traces in the objects of Martina Melilli's film. From the other side, this opacity is asserted as a right, as a site of expression and above all of relation: in Tania El Khoury's performance, our inability to see Basel's body challenges us to overcome a perceptual habit that is also a form of obsessive and violent control, the same habit that reduces and imprisons an individual in their fingerprints or in a stereotype, and simultaneously, absurdly, renders us completely blind, unable to "be with each other" (Fanon 1996).

What role or space of action can a work of art have when it comes to defining ourselves as subjects, as citizens of human beings? To be sure, these two pieces of artwork do not permit us any consolatory or soothing space. Instead, they keep a wound, and a conversation about this wound, open. Perhaps this way of making art—political, reflective and embodied—this "shocking, even traumatic, certainly disturbing" art, which "proposes not the reconfirmation of identity, but its disorientation" (Chambers 2018, p. 14), can offer a 'cure' for the cultural anaesthesia that afflicts our gaze and our relationship with the Other.

Post-Scriptum: On Who Writes, Writing from the Body, and the Body of the Artist or the Intellectual

O my body, make of me always a man who questions! (Franz Fanon, *Peau Noire, Masques Blanches*, Seuil, Paris, 1952)

On 31 May 1975, at Bologna's Modern Art Gallery, the artist Fabio Mauri, during his performance *Intellectual*, used Pier Paolo Pasolini's body like a screen, projecting on it Pasolini's film *Il Vangelo Secondo Matteo*. Sitting in the dark, Pasolini's body becomes a screen, his taut chest literally embodying his own language. Pasolini cannot see his work, he can only know it, or remember it with his body, like the Condemned in Kafka's short story "In the Penal Colony" (1919). In this way the intellectual can feel the full responsibility for this embodiment—a political and ethical responsibility. Regarding this performance, Fabio Mauri wrote that the projection, "physically reveals the birth of the 'intellectual sign' inside the body of the author," but also entails "the imposition of a 'passion' that the author endures, so that he seems to respond physically to what he has conceived" (Alfano Miglietti 2012, p. 27). Being an *Intellectual* (the title of the performance), acting ethically and politically in the world, implies perhaps precisely this 'throwing the body into the struggle'.

References

Agamben, G. 2009. Dell'utilità e degli inconvenienti del vivere fra spettri. In *Nudità*. Roma: Nottetempo.

Alfano Miglietti, F. 2012. *Fabio Mauri. The End*. Milano: Skira.

Appadurai, A. 2011. *Le aspirazioni nutrono la democrazia*. Milano: et al. Edizioni.

Beneduce, R. 2010. *Archeologie del trauma. Un'antropologia del sottosuolo*. Roma: Laterza.

Chambers, I. 2018. *Paesaggi migratori*. Roma: Meltemi.

Connerton, P. 1999. *Come le società ricordano*. Roma: Armando Editore [1989. *How Societies Remember*. Cambridge: Cambridge University Press].

Del Boca, A. 2005. *Italiani brava gente?* Vicenza: Neri Pozza Editore.

Devereux, G. 1984. *Dall'angoscia al metodo nelle scienze del comportamento*. Roma: Istituto dell'Enciclopedica Italiana Treccani [1967. *From Anxiety to Method in the Behavioral Sciences*. Paris: Mouton & Co/The Hague].

Fanon, F. 1996. *Pelle nera, maschere bianche*. Milano: Tropea [1952. *Peau Noire, Masques Blanches*. Paris: Seuil].

Fassin, D. 2016. *Quando i corpi ricordano*. Lecce: Argo [2006. *Quand les corps se souviennent*. Paris: Editions La Découverte].

Grechi, G., and V. Gravano. 2016. Presente imperfetto: una vertiginosa prossimità. In *Presente imperfetto. Eredità coloniali e immaginari razziali contemporanei*, ed. G. Grechi and V. Gravano. Milano: Mimesis.

Gregos, K., and V. Meessen. 2015. *Personne et les autres, Belgian Pavillion catalogue 56ª Venice Biennial*. Milano: Mousse Publishing.

Hall, S., and M. Sealy. 2001. *Different. A Historical Context*. London: Phaidon.

Mauss, M. 1965. *Le tecniche del corpo*. In *Teoria generale della magia e altri saggi*. Torino: Einaudi [1936. Les Techniques du Corps. *Journal of Phsychology* (32): 34].

Mbembe, A. 2005. *Postcolonialismo*. Roma: Meltemi [2000. *On the Postcolony*. Berkeley: University of California Press].

Merleau-Ponty, M. 2002. *Conversazioni radiofoniche fra il 9 ottobre e il 13 novembre 1948*. Milano: SE.

Perec, G. 1994. *L'infraordinario*. Torino: Bollati Boringhieri [1989. *L'infra-ordinaire*. Editions du Seuil].

Ricoeur, P. 2004. *Ricordare, dimenticare, perdonare*. Bologna: Il Mulino.

Rosaldo, R. 1989. *Culture and Truth*. Boston: Beacon Press.

Scheper-Hughes, N. 1994. Embodied Knowledge: Thinking with the Body in Critical Medical Anthropology. In *Assessing Cultural Anthropology*, ed. R. Borofsky. New York: McGraw-Hill.

PART II

Bodies

CHAPTER 6

Carne Nera

P. Khalil Saucier

> We might concede, at the very least, that sticks and bricks *might* break our bones, but words will most certainly *kill* us. (Spillers 1987: 68)

I.

In the fall of 2017 a grainy undercover video appeared on news outlets throughout the world: a video of a modern-day Libyan slave market. "Does anybody need a digger?" the Libyan smuggler inquired in his successful attempt to seduce a buyer of his wares outside the city of Tripoli. An international firestorm ensued. Activists and scholars, known to many as the "new abolitionists," responded to this spectacle with a multitude of progressive anti-racist gestures and activities (Middleberg 2018). Neo-abolitionist analyses outlined the causes for the emergence of twenty-first-century slave markets on the North African coast, while also encouraging

P. K. Saucier (✉)
Africana Studies Department, Bucknell University, Lewisburg, PA, USA
e-mail: pks008@bucknell.edu

G. Proglio et al. (eds.), *The Black Mediterranean*, Mediterranean Perspectives, https://doi.org/10.1007/978-3-030-51391-7_6

people to look at the principles underlying the struggles (namely those led by civil society organizations) taking place against EU border policies and the cathartic calls for the freedom and autonomy of all peoples regardless of citizenship status and/or birthplace (O'Connell Davidson 2017b). The purpose of this chapter, therefore, is to briefly explore the ways in which anti-racist praxis, and more specifically neo-abolitionism in the Mediterranean Basin, creates a crisis that undermines itself, especially as it relates to Black ~~Africans~~[1] on the move.[2]

Despite being stimulated by Black deaths in the Mediterranean and the suffering on the auction blocks of North Africa, neo-abolitionism says little about racism and nothing about anti-Blackness. Much of the discourse is constituted by the language of statist politics, sovereignty, and political economy. Although the global neoliberal order is a major factor in the problem at hand, it cannot fully account for what is at play in Libya and beyond. As this chapter suggests, what we are facing today is a new declination of an old and repressed issue that haunts and composes the European project and modernity itself: anti-Black sociality. This chapter takes seriously the question uttered by the Libyan smuggler about one's need for a *digger*, a labourer known for their capacity to create protective trenches and keepers of sacred vestments. The annihilating call of the smuggler says something beyond the sociological and economic. The digger also serves an analytical function which helps interrogate the ontological grammar that plagues the "neo-abolitionist era" in the Black Mediterranean (Allain and Bales 2012). In other words, the calling out for a digger is a performative announcement of the body made black flesh. The digger is a figure deeply rooted in a structure of feeling that permeates past and present. The digger is Frantz Fanon's "Negro," Sylvia Wynter's NHI, a figure that is more than crestfallen, but damned (Fanon 1967; Wynter 1994).

Put slightly differently, this chapter explores the ways in which neo-abolitionism is (in)vested in the "human" as an all-encompassing and non-racial category for thought and by extension praxis, rather than, as Fanon (1963) forcefully posited, "the damned of the earth" or what I call *carne nera* (Black flesh), which compels us to rethink traditional discourse and

[1] My use of Africans is inspired by Warren (2018). I do this to show how the Africans' blackness overrides their national and ethnic origins.

[2] For the first incisive critique of this problem, see Woods (2013). Of additional support, see Saucier and Woods (2014). Relatedly, see also Danewid (2017).

by extension epistemology.[3] Once *carne nera* becomes the integral unit of analysis for understanding contemporary forms of policing Europe's borders, the cognitive model of the universal human as an important political and paradigmatic articulation, then and now, is thrown into crisis as is antiracist discourse that revolves around issues of "security," "rights," "citizenship," "refugee status," "democracy," and much more. *Carne nera* as a unit of analysis helps to subvert and think differently about the choreography of power in the Mediterranean Basin in order to rectify the shortcomings of abolitionism. In short, this chapter highlights the inability of abolitionism, particularly as it pertains to Black ~~African~~ migration/trafficking, to eradicate anti-Blackness.

By extension, this chapter takes up a Black study of what has been keenly called the Black Mediterranean in order to reconceptualize the grammar of its coherence (Saucier and Woods 2014). Black Mediterranean studies has become somewhat of an academic industry spanning from the ecological and sociological to the poetic and creative. Yet little has been said about the how "our present model of being/definition-of-the-human is given dynamic 'material' existence" (Wynter 1990: 364). That is, little is said about the materiality of violence that incarnates (non)beings. While my comments benefit from the critical, albeit pessimistic, observations about Blackness and modernity from the field of Critical Black Studies, they should not be mistaken as a form of intellectual or praxical resignation. Nor should they be construed as not cultivating insurgent possibilities. Rather, my attempt is to try and chart a brief philosophical course that looks at Black ~~African~~ migrants to Europe while being persistently vigilant of "stay[ing] within the anxiety of [racial] antagonism (Sexton 2016: 4)." In staying with the antagonism that belts the world, we must set aside the empirical for the philosophical and transcendental. In addition, my comments should not be taken as the deanimation of Black social life in Europe, but rather as an exploration of the libidinal and psychic function of the digger within the discursive terrain of the Western imaginary. In other words, my claims are based off of a Black study of migration/trafficking and the continuous enunciation of violence that continues to sustain the regime of immanent distinction.

[3] My use of *carne nera* is no doubt inspired by a reading of Spillers'(1987). Spillers makes the distinction between "body" and "flesh," where black flesh is relegated through forms of violence to the zone of jeopardy and peril.

II.

Scholars and abolitionists Jean Allain and Kevin Bales (2012) suggest that we are currently in the midst of a "neo-abolition era." Many people, from on-the-ground activists to philanthrocapitalists (Chuang 2015), are motivated "to end slavery in fact," which is to say to end the myriad of ways in which "slavery can be [and is] achieved" (Allain and Bales 2012: 502). Modern-day slavery, for the likes of Allain, Bales, and many more abolitionists, is conflated with peonage, trafficking, sex work, child labour, forced labour, and much more, for all lead to "control tantamount to possession (Allain and Bales 2012: 510)." As Janie Chuang has astutely observed, the discursive acrobatics of abolitionism often conflate forced labour with trafficking and trafficking with slavery (Chuang 2014). In other words, modern-day slavery is understood as an activity, rather than as a regime of social organization (Saucier and Woods 2014; Woods 2013). Given the capacious and diverse forms of activity *as* modern-day slavery, abolitionism is comprised of variable practices, analytical capacities, and desires that are beyond the scope of this chapter.[4] Nonetheless, there is a remarkable consistency in (1) their emphasis on the political economy and (2) their politics of race.

Political Economy

Modern-day trafficking/slavery in the Mediterranean Basin is most often understood as a political economic problem; a "nasty business" derivative of a myriad of global political and economic forces (Allain and Bales 2012: 510). As Jared Sexton (2015a) points out in his short yet cogent essay "Don't call it a comeback," there are roughly two sets of abolitionists: first, those who understand modern-day slavery from within the infrastructures of neoliberalism, included but not limited to supply chains, development projects, immigration policies, and consumption practices; and second, those that lay diagnostic blame on deviant and culturally bankrupt entrepreneurs that profit from political instability, government austerity, and economic precarity.[5] It is this explanation that has gripped

[4] Well beyond the scope of this chapter, for example, is the distinction between prohibitionist (prohibiting) and abolitionist approaches. See Nagel (2015).

[5] One of the main problems with using the Libyan example or any example from continental Africa is the assertion of African culpability, or what we can call the 'cultural void thesis': if Africans are enslaving Africans, it is not a race problem. By extension, Africa would once

civil society and its culture industry's understanding and imagination of the explosion of slave markets in post-Qaddafi Libya.[6]

The work of Sayak Valencia, for instance, is suggestive of a composite approach that understands that "extreme violence…death and thanatopolitics have been converted into a new form of fierce and unapologetic capitalism (2018: 83)." For Valencia, "violence has become the law of the market," and the violence once reserved for organized crime and states is now available to all.[7] This observation, as I will discuss below, can only be made by presupposing a universal subject that excludes Blackness (Valencia 2018: 62). She suggests that "gore capitalism," as she calls it, "*could…be* understood as part of an intercontinental struggle of extreme postcolonialism (Valencia 2018: 76)." More specifically, the trafficker caught on video selling his wares in the Tripoli marketplace is what Valencia understands as the emergence of a new class formation, the "gore proletariat (2018: 160)." This new class formation, working for itself, has "made violence into another market niche, recontextualizing the position of the body as a strategy for accumulation (Valencia 2018: 161)." "Violence," she continues "is converted into a resource for the gangster to manage, produce, and sell; it has become the tool sine qua non to carve out a space on the capitalist ladder (Valencia 2018: 197)." Even someone like Valencia (2018), whose work is at once considered decolonial, feminist, and working at the critical edge of abolitionism, fails to extend her analysis beyond the production of capital. She tangentially discusses race but does so only through the rubric of colonialism, for example the British occupations in Nigeria, India, and elsewhere. Her use of violence becomes a non-racial extension of the global neoliberal order similar to the ways in which sex trafficking, human trafficking, and more are often understood from a non-racial ahistorical lens (Woods 2013). She does not illustrate how gore capitalism might be nothing more than the blackening of third world

again be the land of savagery. See Woods 2013 for an extended mediation on deviance understood as the result of "African culture."

[6] See for example, https://www.reuters.com/article/us-libya-security-rights/executions-torture-and-slave-markets-persist-in-libya-u-n-idUSKBN1GX1JY; http://time.com/5042560/libya-slave-trade/; https://www.newsweek.com/humans-sale-libyan-slave-trade-continues-while-militants-kill-and-torture-855118; Mafu, Lucas. "The Libyan/Trans-Mediterranean Slave Trade, the African Union, and the Failure of Human Morality." *SAGE Open* 9.1 (2019): 2158244019828849.

[7] Of note is the use of "become" which signals a coming into existence, a developing ontology that is arriving on the horizon.

non-Blacks ensnared in the web of trafficking qua slavery, and that the gratuitous violence that produces the Black, past and present, as Achille Mbembe recently observed in *Critique de la raison nègre* (*A Critique of Black Reason*) (2013) works at a different register and thus requires a different analytical application.[8] Put slightly differently, trafficking/migration out of Africa is always already violent and not the result of a failed state and new stage of capitalism, but a technology of ontological policing that fuses to the body politic of human. Slave auctions on the North African Mediterranean coast are simply a warranted part of the social order.

As Tryon Woods has suggested, "[a]s the political economy has changed with time, the symbolic economy of antiblackness persists (2014)." The theory and praxis of political economy has limits. In other words, even in its most radical form, abolitionism. The digger, as I suggest below, marks a political ontology that delimits which humans can lay claim to full human status and which cannot. What the digger has to offer is his/her Blackness, difference on a world scale. The digger is productive even before they are put to work, and work is already a reaction to that productivity, a vicious and violent one. To this end, abolitionism could offer a broader, more dynamic account and honest assessment of modern-day slavery in the Mediterranean Basin beyond political economy. Yet, it continues to make "grave analytical and political error[s]" (Woods 2014) that prioritize the political economy and by extension presuppose a genre of human that is singular, yet comprised of a kaleidoscope of colours, thusly extending an epistemic vice that continues to be structural and paradigmatic.

Politics of Race

According to James B. Stewart, abolitionist discourse suffers from "racial blindness" and "[p]roperly understood, African American history can help to illuminate the problem of slavery today (2015)." Yet at his urging we are bombarded with historical examples such as the convict-lease system in the US and other spectacles of misery, pain, and death that look similar to the practices of slavery today. Once again, what we get is slavery as practice, not a regime of social organization. While scholars and activists mine the past for historical connections, they often do so without taking into account the deep structural inflections of the past on the present, a

[8] As Vargas states, "antiblack technologies are required to continuously expand, they inevitably interpellate nonblacks (but potentially soon to be) black bodies (2016: 67)."

temporality understood by Hartman as the "time of slavery (2002)." This corrective often misses the point, for Blackness becomes merely a historical referent and/or empirical supplement, rather than the source and sequence of modernity.

Julia O'Connell Davidson is perceptive to ask us to "draw on history" but resist the temptation to create false analogies of past and present, analogies that obscure the living past (2017a: 5). O'Connell Davidson states, "The idea of race as constitutive of borders between flesh and blood human beings in terms of their capacities, moral worth and rights is inexplicably bound to the history of Atlantic world slavery (2017a: 5)." For O'Connell Davidson, neoabolitionism "actively deflects attention" from the centrality of race and "works to minimise the scale and *nature* of the atrocity of transatlantic racial slavery (2017a: 5, emphasis added)." Similarly, Joel Quirk notes:

> Whenever "modern-day abolitionists" look backwards into the past, they tend to selectively focus upon the history of anti-slavery activism in Britain and the United States. Within this context, the history of abolition has been chiefly approached as a source of instruction and inspiration. When it comes to instruction, the main focus has been the tactics and techniques used by mostly white Anglo-Saxon pioneers, such as petitions, novels, publications, networks, boycotts, lawsuits, meetings, and artistic icons. When it comes to inspiration, the early history of anti-slavery has been celebrated as a key illustration of the power of ethical leadership, collective action, and personal commitment. (2015)

In other words, the history of racial slavery is incorporated in order to (1) construct an alibi for dismissing what is "constitutive of borders between flesh and blood human beings," that is as a way to flatten any and all difference in political ontology; and (2) as a way to bolster the image of a knowledgeable and compassionate abolitionism. The disavowal of ongoing reformation of racist practices and one's benefit from it is a rehumanizing effort, not of the Black, but of the activist. Race as history and identity is incorporated, but abolitionism elides the "extravagant abjectness" of Blackness (Broeck 2018; Scott 2010).

Even the critical and important work of Kamala Kempadoo fails to attend to the centrality of Blackness, choosing instead to structure her arguments around the normative racial binary of "white/other." In her work on contemporary campaigns against trafficking and modern-day slavery she states that we are witnessing

> a neoliberal white chivalrous crusade across the world, born of a moral sense of goodness that shores up the power and subjectivity of the North, with the 'developing' Global South and East as the dumping grounds for helping imperatives involving rescue and charity. They have little effect on the causes of the problem and the subjectivity and humanity of the Other is secondary. (Kempadoo 2016)

Kempadoo then evokes Fanon's "wretched," but says nothing about how "wretchedness" operates beyond a generalized descriptive category. "Wretched," for Kempadoo (2016), is the multiracial Other fleeced in Whiteness as burden despite having suggested how said campaigns are subject-making events.

Abolitionists are right to engage race. However, what is missing is an awareness that understands Blackness functions differently from otherness, especially since otherness is not nonbeing in the way Blackness is. O'Connell Davidson, Quirk, and others are correct in asserting that racial slavery be part of any and all understanding of trafficking/modern-day slavery, for its residual effects can be found in all language, memory, and subjectivity. As Christina Sharpe limns in *Monstrous Intimacies*, "all modern subjects are post-slavery subjects (2009: 3)." Once again, the ideas and histories of Blackness are used not for its own liberation, but as a way in which to recuperate liberal pluralism that undergirds abolitionist discourse. Abolitionism in this sense is very much cut from the same political cloth as Jürgen Habermas's conception of modernity as an "unfinished project (1992)." Rather than a critique of modernity and its condition of possibility in and through Blackness, neo-abolitionism is often in support of modernity. Therefore, the error in liberal thought is ignoring the problem of universality, a universality that takes for granted the inclusion of all sentient beings (Da Silva 2007; Mills 2017; Wynter 2018). Needless to say, abolitionism does not capture racial slavery as paradigm.

Back to the grainy video of the Tripoli slave market. I believe it provides a moment to (re)focus on the blindness that plagues abolitionism: Blackness as antagonism to apply pressure to the decadent coherence of liberalism and humanism.[9] Whether explicit or not, the cognitive mechanisms of neo-abolitionism are rooted in liberal Western humanism, which, according to many from within the Black radical tradition such as Aime Césaire, Sylvia Wynter, Lewis R. Gordon, and many more fail to understand that race contains an anthropological rubric for the human; the anti-human is the yardstick (Césaire 2001; Gordon 2013; Wynter 1984; Wynter and Scott 2000). The digger helps define the contours of humanity: the boundaries of what is considered human, and what is at its borders (the not-quite-human) or completely outside its conception (the non-human). Any attempt to theorize race without placing centrally the lived experiences of those precisely defined as not-quite and non-human re-enacts a racializing violence. In other words, the doxastic attitude within abolitionist circles and anti-racist circles more generally assumes that anti-Blackness is merely consciousness and can be rejected by simply assuming one's humanity,. The presupposition of granting humanity is not simply and merely a form of consciousness, or by making historical and empirical observations analogizing past and present. Joao Costa Vargas understands this as a form of "oblique identification" when "Black processes are recognized only partially, belatedly, indirectly, reluctantly, or even unknowingly" (2018: 5)—or, similarly, the instrumentalization of Blackness for any and all purposes except for itself. As Saidiya V. Hartman remarks, the "language of freedom no longer becomes that which rescues the slave from his or her former condition, but the site of the re-elaboration of that condition" (Hartman and Wilderson 2003: 185). Black suffering and death is politically and existentially leveraged in an attempt to reconfigure the human tainted by the dastardly actions of the gore proletariat.

Thus, the provocation is simple: what if a universal subject is not very useful for thinking through Black migration, trafficking, labour exploitation, and much more? What if the Blackness of the digger demands a different set of agreements? According to Jared Sexton, what happens when "the concept of the human, or a certain dominant conception of the human, stands in the way…?" (2015b: 163).

[9] One might be hard-pressed to state that anti-trafficking scholars and activists seldom talk of race and racism. Any quick glance at the *Anti-Trafficking Review*, for example, suggests that race is central, but not Blackness.

III.

Is the annihilating call of the seller any different from Frantz Fanon's "Look, a Negro?" (1967: 89). Working in and from the negative of non-being found in *Black Skin, White Mask* (1967) and *The Wretched of the Earth* (1963), the digger is also a symbol and sign of the very thing that belts the world: anti-Blackness. The call for a digger is catalyzed by the interface of past and present. Thus, the selling of a digger disallows any reading of racial terror as derivative of modernity, but a forewarning of the violence to come. The call for a digger tells us something about the essential predication of the Black on the auction block. The existence of classificatory categories such as trafficker and migrant are logically independent of ontological classification.[10] The digger *is* part of a choreography of power between the human and non-human. Thus, my argument is not platonic: rather than coming from pure ideas, it is derived from a materiality of violence that incarnates (non)beings. And the "digger" suggests as much. The digger *is* being-named, it subjects the Black to a violence that exceeds any analogy despite many non-Blacks being sold in the Tripoli market and elsewhere. (Again, why not a moment of others being Blackened?) Additionally, my argument shows racial slavery in its complete form, exceeds labour and possession. This conversion creates a scene in which the Black ~~African~~ is bound to a violence that is inseparable from non-being: in other words, a linguistic utterance, a statement of kind that locates Fanon's "zone of non-being (Fanon 1967; Gordon 2005)." The digger signals more than the deviance of individuals and gruesome reality of outrageous labour conditions; it also, and more importantly, signals the "violence that turns a body into flesh, ripped apart literally and imaginatively (Wilderson 2010: 38)." I do not want to confine the digger to a sociological category, but rather present it as a marker for a political ontology, a figure deeply rooted in a structure of feeling that permeates the present. In this sense I am following Ronald Judy's observations that "'the negro,'" or in this case the digger, "is not empirical but entirely discursive or transcendent…the sort of knowledge which has no empirical content" (1991: 15). Put slightly differently, Blacks are no doubt *anthropos*, but they are not afforded entry into *humanitas* (Paris 2018).

Returning to the universality that plagues trafficking/modern-day slavery discourse, including that which is attentive to race, we confront the

[10] I am not arguing what is *in* a subject, but what is said *of* a subject.

tension between "equally existing" and "existing equally:" an abstraction neglected in abolitionist and anti-racist circles. For example, the Syrian migrant and the Black from Nigeria equally exist, but their existence is not equal, for one is imparted a humanity that is withheld from the other. Mere existence does not assure inclusion into humanity, and existence gives no right to the constitution of a human space. At best it gives the right to be included into a space always already constituted by others. Humanity, therefore, is an attribute granted by inclusion, whereas sentience is a function of birth. This difference in *types* of beings can be explored through the flesh of the digger and the body of the human.

My use of *carne nera* implores that is a reading of Hortense Spillers where the continuum of violence and theft of the body locate ground zero for the conceptualization of the human. While Spillers' work may be overused in some quarters of the academy, her distinction between flesh and body offers an important intervention for Black Mediterranean studies more generally and abolitionist circles more specifically, for all flesh is common until it is marked, and the marking of flesh is accomplished by violence. For Spillers, racial slavery is initiated by a literal and figurative theft of the body (Spillers 1987: 68). She highlights the zone where power transforms bodies into flesh drawing a "cunning difference" (1984:76): The body is the sphere of subjectivity situated by liberal notions of freedom, while flesh is the "primary narrative" of captivity, the mere annex of another (Spillers 1987: 67).[11] Thus, the black body remains captive as flesh through the modern-day auction block and much more. Black bodies remain exposed, despite the profundity of the Black Mediterranean, and even as individuals win their right to their own bodies. According to Calvin Warren, a perceptive reader of Spillers, Black being is "pure function and not relation…[that is] to inhabit this treacherous position of function, which enables human beingness to engage in its projectionality." (Warren 2018: 32). Therefore, why are we to think anything will be different after the abolition of modern-day slavery?

The Black, not the Syrian for example, torments the metaphysical stability of Europe, for Europe needs the Black but is allergic to its existence. In this sense, the digger helps (re)construct the boundaries of the human, "the constitutive figuration of Euro-American modernity" (Broeck 2018: 17). Thus when only one or more than a thousand move across the aquatic threshold of the Mediterranean, we are witnessing the drift of boundaries

[11] See also Da Silva (2009).

between the human and non-human. The digger is not just advertised to an economic world, but also a metaphysical world. The political predication of Black being is pure function. Again, as Warren states clearly and without philosophical doubt, the "black being's function within metaphysics is to inhabit the void of relationality" to "help the human being" (2018: 32). Similar to the moats and trenches that diggers are known to create, *carne nera* is "form and formlessness, property and human" (Warren 2018: 36–37) or "the passageway that encompasses the terrain between the living and the dead" (Holland 2000: 43). The digger remains inert and neutralized beyond the slave market. The digger helps illustrate the ways in which Blackness is still open and available for circulation. In fact, the digger could not even speak of his labour expertise; he was spoken for. Seeing the Black ~~African~~ on the modern-day auction block captures the vestibule of the absence presence of Blackness that so often plagues neo-abolitionist practices (Spillers 1987). In the end, to see the digger strictly and only in economic terms displaces the import of racial slavery "as [the] nomoi of modern politics (Weheliye 2014: 36)." The digger is plagued by a stigmata of distinction beyond identity set in motion long ago. Even if the digger gains entry into humanity, the "hieroglyphics of the flesh" do not vanish or wither away(Spillers 1987: 67).

If the human is thought to be transhistorical, it therefore sets the terms for how freedom and liberation are thought, represented, and fought for. By extension, knowingly or not, neo-abolitionism conceals an ontological project of devastating distinction. *Carne nera*, alternatively, challenges us to think beyond the human and position Blackness at the vortex of any and all freedom movements: the Black radical tradition teaches us how to (re)present freedom beyond liberal humanisms addiction to rights and sovereignty. As Warren has observed, "*Black rights would be the end of human rights*" (2018: 85). Or as Vargas argues the "incapacity to engage structural antiblackness is tantamount to the denial of antiblackness" (2018: 241). In other words, Black flesh is not simply and only the point of departure, but the vortex of struggle. If Black flesh is merely the point of emphatic departure, then we simply perform "the multi-racial project of the empire-state" (2018: 258). Understanding from the position of Black flesh extends beyond redemption of the Western world and its inclusive ideals, bringing us closer to that moment in *Black Skin, White Mask* where "utter declivity" births "an authentic upheaval (Fanon 1967: 8)." *Carne nera* is the ground from which "guerrilla intellectuals" (Rodney

1990) have been and are born and from which the "mechanics of rebellion" (Wynter 1979) are honed. *Carne nera* is the ontological totality of something else: *being* as *carne nera*.

IV.

Carne nera ethically demands a critique in neo-abolitionist and anti-racist formations. In this chapter I have tried to briefly sketch the theoretical presuppositions about the human and what a Black study might offer in terms of thinking differently, and more radically, about the choreography of power taking place in the Mediterranean. It is my understanding that *carne nera* is not as "a category but as a referent of another mode of existing in the world…fractures the glassy walls of *universality*" (Da Silva 2017). According to the Black radical tradition, neo-abolitionists must take stock of their position, omissions, and performative solidarities and think about what is really required to curtail anti-Blackness. We need to privilege more than the margins; rather, we need to think about how to approach the Black Mediterranean in a way that does not reproduce the oppressions we are trying to unmask. In short, *carne nera* is the counterforce and counter-memory that ruptures the libidinal economy. It is the ultimate sublimation.

References

Allain, J., and K. Bales. 2012. Slavery and Its Definition. *Global Dialogue* 14 (2): 6–14.

Broeck, S. 2018. *Gender and the Abjection of Blackness*. Albany: SUNY Press.

Césaire, Aimé. 2001. *Discourse on Colonialism*. New York: New York University Press.

Chuang, J.A. 2014. Exploitation Creep and the Unmaking of Human Trafficking Law. *American Journal of International Law* 108 (4): 609–649.

———. 2015. Giving as Governance: Philanthrocapitalism and Modern-Day Slavery Abolitionism. *UCLA Law Review* 62: 1516.

Da Silva, D.F. 2007. *Toward a Global Idea of Race*. Vol. 27. Minneapolis: University of Minnesota Press.

———. 2009. No-Bodies: Law, Raciality and Violence. *Griffith Law Review* 18 (2): 212–236.

———. 2017. 1 (Life)÷ 0 (Blackness)=∞-∞ or∞/∞: On Matter Beyond the Equation of Value. *Re-visiones* (7) pp. 1–12.

Danewid, I. 2017. White Innocence in the Black Mediterranean: Hospitality and the Erasure of History. *Third World Quarterly* 38 (7): 1674–1689.

Fanon, F. 1963. *The Wretched of the Earth*. New York: Grove Press.

———. 1967. *Black Skin, White Masks*. New York: Grove Press.

Gordon, L.R. 2005. Through the Zone of Nonbeing a Reading of Black Skin, White Masks in Celebration of Fanon's Eightieth Birthday. *The CLR James Journal* 11 (1): 1–43.

———. 2013. *Existentia Africana: Understanding Africana Existential Thought*. New York: Routledge.

Habermas, J. 1992. *Cultural-Political Interventions in the Unfinished Project of Enlightenment*. Cambridge, MA: MIT Press.

Hartman, S.V. 2002. The Time of Slavery. *The South Atlantic Quarterly* 101 (4): 757–777.

Hartman, S.V., and F.B. Wilderson. 2003. The Position of the Unthought. *Qui Parle* 13 (2): 183–201.

Holland, S.P. 2000. *Raising the Dead: Readings of Death and (Black) Subjectivity*. Durham: Duke University Press.

Judy, R. 1991. Kant and the Negro. *Surfaces* 1 (8): 4–70.

Kempadoo, K. 2016. Revitalizing Imperialism: Contemporary Campaigns Against Sex Trafficking and Modern Slavery. *Cadernos Pagu* (47).

Mbembe, A. 2013. *Critique de la raison nègre*. Paris: La Découverte.

Middleberg, M. 2018. How to Stop the Slave Trade in Libya and Beyond. *The CNN Freedom Project*, January 5. Available at https://www.cnn.com/2018/01/05/opinions/maurice-middleberg-stopping-the-slave-trade/index.html. Accessed 5 July 2019.

Mills, C.W. 2017. *Black Rights/White Wrongs: The Critique of Racial Liberalism*. New York: Oxford University Press.

Nagel, M. 2015. Trafficking with Abolitionism. An Examination of Anti-Slavery Discourses. *Champ Pénal/Penal Field* (12).

O'Connell Davidson, J. 2017a. The Presence of the Past: Lessons of History for Anti-Trafficking Work. *Anti-Trafficking Review* (9): 1–12.

———. 2017b. Libyan Outrage: Slavery or Borders? *OpenDemocracy*, December 12. Available at https://www.opendemocracy.net/en/beyond-traffickingand-slavery/libyan-outrage-slavery-or-borders/. Accessed 25 June 2019.

Paris, W.M. 2018. Humanism's Secret Shadow: The Construction of Black Gender/Sexuality in Frantz Fanon and Hortense Spillers. *philoSOPHIA 8* (1): 81–99.

Quirk, J. 2015. Uncomfortable Silences: Anti-Slavery, Colonialism, and Imperialism. *OpenDemocracy*, October 22. Available at https://www.opendemocracy.net/beyondslavery/joel-quirk/uncomfortable-silences-anti-slavery-colonialism-and-imperialism. Accessed 6 Apr 2019.

Rodney, W. 1990. *Walter Rodney Speaks: The Making of an African Intellectual.* Trenton: Africa World Press.

Saucier, P.K., and T.P. Woods. 2014. Ex Aqua: The Mediterranean Basin, Africans on the Move, and the Politics of Policing. *Theoria* 61 (141): 55–75.

Scott, Darieck. 2010. *Extravagant Abjection: Blackness, Power, and Sexuality in the African American Literary Imagination.* New York: New York University Press.

Sexton, J. 2015a. Don't Call It a Comeback: Racial Slavery Is Not Yet Abolished. *OpenDemocracy*, June 17. Available at https://www.opendemocracy.net/en/beyond-trafficking-and-slavery/dont-call-it-comeback-racial-slavery-is-not-yet-abolished/. Accessed 10 Apr 2019.

Sexton, Jared. 2015b. Unbearable Blackness. *Cultural Critique* 90 (Spring): 159–178.

———. 2016. Afro-Pessimism: The Unclear Word. *Rhizomes: Cultural Studies in Emerging Knowledge* 29: 29.

Sharpe, Christina. 2009. *Monstrous Intimacies: Making Post-Slavery Subjects.* Durham: Duke University Press.

Spillers, H. 1984. Interstices: A Small Drama of Words. In *Pleasure and Danger: Exploring Female Sexuality*, ed. C.S. Vance, London: Routledge 73–100.

Spillers, H.J. 1987. Mama's Baby, Papa's Maybe: An American Grammar Book. *Diacritics 17* (2): 65–81.

Stewart, J. 2015. The New Abolitionists and the Problem of Race. *OpenDemocracy*, April 21. Available at https://www.opendemocracy.net/beyondslavery/james-brewer-stewart/%E2%80%98new-abolitionists%E2%80%99-and-problem-of-race. Accessed 12 June 2019.

Valencia, S. 2018. *Gore Capitalism.* Vol. 24. Cambridge, MA: MIT Press.

Vargas, J.C. 2016. Can the Line Move? Antiblackness and a Diasporic Logic of Forced Social Epidermalization. In *Critical Ethnic Studies: A Reader*, ed. N. Elia, D.M. Hernández, J. Kim, S.L. Redmond, D. Rodríguez, and S.E. See, 63–91. Durham: Duke University Press.

———. 2018. *The Denial of Antiblackness: Multiracial Redemption and Black Suffering.* Minneapolis: University of Minnesota Press.

Warren, Calvin. 2018. Ontological Terror: Blackness, Nihilism and Emancipation. Duke University Press.

Weheliye, A.G. 2014. *Habeas Viscus: Racializing Assemblages, Biopolitics, and Black Feminist Theories of the Human.* Durham: Duke University Press.

Wilderson, F.B., III. 2010. *Red, White & Black: Cinema and the Structure of US Antagonisms.* Durham: Duke University Press.

Woods, T.P. 2013. Surrogate Selves: Notes on Anti-Trafficking and Anti-Blackness. *Social Identities* 19 (1): 120–134.

———. 2014. The Antiblackness of Modern-Day Slavery Abolitionism. *OpenDemocracy*, October 10. Available at https://www.opendemocracy.net/

en/beyond-trafficking-and-slavery/antiblackness-of-modernday-slavery-abolitionism/. Accessed 1 Feb 2019.

Wynter, S. 1979. Sambos and Minstrels. *Social Text* 1: 149–156.

———. 1984. The Ceremony Must Be Found: After Humanism. *Boundary* 2: 19–70.

———. 1990. Beyond Miranda's Meanings: Un/silencing the 'Demonic Ground' of Caliban's 'Woman'. In *Out of the Kumbla: Caribbean Women and Literature*, ed. C.B. Davies, 355–372. Trenton: Africa World Press.

———. 1994. No Humans Involved: An Open Letter to My Colleagues. In *Forum NHI: Knowledge for the 21st Century*, vol. 1, no. 1, 42–73. Stanford: Institute NHI.

———. 2018. Beyond Liberal and Marxist Leninist Feminisms: Towards an Autonomous Frame of Reference. *The CLR James Journal* 24 (1/2): 31–56.

Wynter, S., and D. Scott. 2000. The Re-Enchantment of Humanism: An Interview with Sylvia Wynter. *Small Axe* 8: 119–207.

CHAPTER 7

Impermanent Territories: The Mediterranean Crisis and the (Re-)production of the Black Subject

Timothy Raeymaekers

I would like to thank the Swiss Network of International Studies (SNIS) for kindly granting me the opportunity to engage in this research through my directorship of the project "New Plantations: Migrant Mobility, 'Illegality' and Racialisation in European Agricultural Labour" (2015–2017). Furthermore I am extremely grateful to Hervé Papa Latyr Faye, Mbaye Ndiaye, Lamin Faye, Nadine, Omar, Pauline and the other inhabitants of Casa Sankara for their support. Finally I would like to thank my colleagues at the Political Geography unit, particularly Muriel Côte, Rory Rowan, Rony Emmenegger and Benedikt Korf, for their continuous suggestions and constructive critique. I have presented various strands of this research at conferences and workshops in intense communication with the Black Mediterranean collective, with whom I present this issue and to whom I am endowed with gratitude. All the errors and misinterpretations in this chapter are obviously completely my own.

T. Raeymaekers (✉)
Department of Geography, University of Zurich, Zürich, Switzerland
e-mail: timothy.raeymaekers@geo.uzh.ch

G. Proglio et al. (eds.), *The Black Mediterranean*, Mediterranean Perspectives, https://doi.org/10.1007/978-3-030-51391-7_7

> Look at what they've accomplished, the leaders of Europe: *La nouvelle Libie* (shrugs). They've chased us Africans away from the continent… Now they lament that too many migrants are arriving at their doorsteps. But who has created this problem, this chaos in the first place? It is them! (Omar, Casa Sankara, 28 July 2018)

> This place is worse than a prison… At least there you know what to expect, what is your crime and what is your punishment. Here we are only dying, slowly, in silence. (Senegalese man at Casa Sankara)

> Man is a mixed up creature. He becomes all the more confused, we might add, the more he is subjected to tensions: at that point he evades our judgment, just as a compass goes wild at the magnetic pole. (Primo Levi: The Grey Zone)

Introduction: The Mediterranean as a Grey Space

This chapter is concerned with the role and function of Black African fugitive sites in the context of a rapidly changing European border regime since the outbreak of the Libyan war in 2011. A new form of settlement has been on the rise in the Mediterranean, which serves both as a transit place and precarious shelter for migrants on the move. Such 'ghettos' acquire different forms, from concrete buildings to plastic shelters to entire migrant villages. They have been spotted on major migrant travel routes in Africa—in Niger, Mauritania, Libya, Tunisia—as well as in Europe, in urban peripheries as well as rural environments. While not a lot of literature exists on these places, they are said to be used as stopover places for the destitute and the weak: those migrants driven by expulsion, violence and loss (Lecadet 2013; Agier and Lecadet 2014; Lucht 2017). At the same time, though, they also serve to forge social networks for supporting onward journeys, to look for jobs, to find financial assistance, and in general to reconstitute a sense of community in a context of high risk and unpredictability.

More and more, therefore, such migrant 'ghettos' have become a fixed sight across the Mediterranean region as places "from where a collective life can be rebuilt" (Lecadet 2014: 157). Since 2004 in Tripoli, Libya, for instance, Sylvie Bredeloup and Mahamadou Zongo (2005) write how Burkinabe migrants who, for the most part have been active in the construction sector there, have started to organize their accommodation around certain *foyers* (literally: households), or *enclos* (enclosures) in the urban periphery, where they self-organize food acquisition, rent and communal life. Their Anglophone companions predominantly used the term 'ghetto' for these living arrangements. Since the outbreak of the war in Libya and the ensuing refugee crisis that resulted from it, this form of precarious urbanization has rapidly spread across the European continent in the form of "informal refugee settlements" (Ticktin 2016; Davies et al. 2017), first in Italy (where they have already existed since the mid-1990s), then in France, Germany, and other places.

Specifically in southern Italy, which will be the focus of this chapter's contribution, the simultaneous intensification of refugee flows and the precarization of migrant lives during the last five years has significantly impacted on the growth of such migrant ghettos. The ghettos are used as refuges for migrants on the way to other destinations. Increasingly, they also serve as permanent clusters for migrant workers who are actively employed in the agrarian economy. One can argue that the confluence of these two factors—increasing refugee flows and increasingly precarious agricultural labour across the South of Europe—has led to an emerging urban-rural *frontier*: a new, peri-urban temporary settlement, which, apart from the creative socio-spatial organization it projects, can also be seen as an interstitial, or a liminal space (Korf and Raeymaekers 2013; Thomassen 2014) where novel forms of politics emerge. Migrant ghettos can thus be described as impermanent territories: they reinforce at once the displacement of marginalized subjects in this fluctuating frontier, while at the same time they constitute an attempt towards new forms of rootedness through webs of dependencies and relationships.

Rather than seeing this phenomenon exclusively in connection to border surveillance,[1] it has to be emphasized that the intensified refugee flows across the Mediterranean since 2011 are also grafting themselves upon already existing migrant settlements, particularly in the South of Europe, where a historical movement of labour migration has left important social and economic marks since the mid-1980s. Particularly in regions like Calabria, Puglia, Sicily and Basilicata (Italy), Kalamata (Greece), Andalusia (Spain) and Torbali (Turkey), resident migrant workers have started to intermingle with newly arriving refugee populations from West and North

[1] In the context of the EU border regime, Papadopoulos, Stephenson and Tsianos (2008: 74) understand this liminal space to be "a flexible regime of control which attempts to regulate mobility flows by forging contingent border zones wherever the routes of migration make the existing regime porous" (Papadopoulos et al. 2008: 74; see also Kasparek 2016).

Africa, the Middle East and East Asia. This has led to new forms of ethnic and urban segregation, whereby African migrants, predominantly but not exclusively from West African origin, end up living in such marginal sites: in informal camp sites and precarious shelters, Malian, Ivorian, Ghanaian and Burkinabe workers in the fields remain subject to severe labour exploitation, deprived of resources and constantly under the threat of eviction (Corrado et al. 2016).

Widening our perspective from a purely humanitarian concern towards an interrogation of the geographic and political transformation that Mediterranean mobilities are potentially generating nowadays, however, we could ask ourselves what kind of life this recent process of informal cross-border urbanization represents. To quote urban theorist Oren Yiftachel (2009), segregated settlements may indeed become an active way to place populations that are not deemed worthy of civilized government between the 'lightness' of legality, safety and full membership, and the 'darkness' of eviction, destruction and death. It is exactly in this liminal space that my main questions for this chapter need to be situated. In this context, I imagine migrant ghettos as part of a widening grey zone, which is metaphorically situated between the rural and the urban, the border and the checkpoint, the 'civilized' arenas of democratic liberal democratic government and the marginalized locations of exploitation, exclusion and repression. So what forms of socio-spatial organization is the intensification of human mobility producing in the context of our contemporary era? And how is this social space constitutive of an emerging, Black political subjectivity? For the purpose of this volume, my questions build both on existing debates in critical urban studies, specifically on the ways in which urbanization deals with the "unplannable" (Roy 2005: 147) phenomena of informal settlement, labour and politics in a context of rapidly shifting and contrasting technologies of government. Furthermore, I also address a distinctively geographic question in this chapter about the specific modes of ordering that aim to govern human mobility *across* space (Brenner and Katsikis 2013).

Between the precarious, informal shelters of travelling migrants, the historical labour settlements and the official reception and detention centres, I analyse what kind of political apparatus is being put in place so as to decide over the rights of populations that are deliberately placed out of sight of the 'normal' democratic decision-making process. Building on this tension between visibility and invisibility of migrant subjectivity, this chapter proposes a new imaginary of the Mediterranean as an expanding grey space. This space is characterized, on the one hand, by active spatial segregation, diffusion and marginalization of Black African migrants across this new urban-rural frontier. On the other hand, it also constitutes the foundation of a new, hybrid politics that promotes the diffuse negotiation of migrant rights in a context of the active reconfiguration of territorial borders (Giordano 2008; Saucier and Woods 2014).

In the next three sections, I will subsequently give an overview of my theoretical and empirical argument. In order to proceed, however, I need to dedicate a few words as to how I place my research in the context of the Black Mediterranean, which has recently re-emerged as a concept to investigate the bordered lives of Africans in southern Europe. Personally, I started using the term back in 2015, when I observed a new Black consciousness emerging in Italy in the context of Europe's rapidly shifting borderscape. On the one hand, this volume gives me the opportunity to place my analysis in a longer reflection on the Mediterranean as an explicitly postcolonial borderland (Chambers 2008; Giaccara and Minca 2010; Lombardi-Diop and Romeo 2015) in which cultural identities are challenged, upset and mixed. Along with others, I have observed how intensifying migration flows across the Mediterranean waters since 2011 have generated an open crisis that also motivates Africans to claim a proper place where their rights are actively being denied and trespassed (Raeymaekers 2014; Hawthorne 2017; Merrill 2018).[2] On the other hand, I also emphasize the explicit racial geographies of these engagements in the context of a changing topology of border and migration governance. The Libyan conflict has been said to be opening the flood gates of outward African migration, a threat that was once associated to the late colonel Muammar Ghadaffi, but is now becoming reality through the active trafficking of migrants by Libyan armed groups (Brambilla 2014).

[2] A tipping point in this respect has been the conflicting memory of the Lampedusa tragedy, whereby 386 Eritrean, Somali and Ghanaian migrants died on board a capsized ship just ashore the island of Sicily in 2013. The flagrant denial of humanitarian assistance has not only raised a shockwave throughout Africa and Europe, but also formed the backdrop to new political connections. A mixed network of (White European) border activists, Black African youngsters and various non-state organizations have started to uneasily cling together in a joint public space, activating sentiments of common precarity and political consciousness, forming a mixed and contentious community of reinvented Leftist radicalism, Black rights movements and ethnic nationalism. In several cities, these contestations have led to physical occupations of abandoned and derelict buildings, former public and private infrastructures that have been wielding the Lampedusa banner as a sign of solidarity and companionship with those who rightlessly died on the way.

At the same time, the proceeding privatization of the Mediterranean border space through externalized camp sites, digitalization and tracking techniques and the operation of privatized bordering agencies (including militias and armed actors) is making European outsourcing states increasingly complicit in the mistreatment of migrants and asylum claimants who come to Europe to seek protection. Frequently, the latter end up being systematically abused. Particularly Black African migrants—who have constituted the majority of asylum claims in Italy since 2011—are frequently caught in the middle between wanting to be seen by state bureaucracies that ideally should grant them protection, and a desire to improve their livelihoods through often risky and fragmented trajectories (see also Borri 2017; Fontanari 2017; Pinelli 2017).

To emphasize this political and geographic dimension of the Black Mediterranean, I focus my attention on the place-making aspects of today's migrant ghettos in the Mediterranean. Following the observation that we indeed need to consider these sites as forming part of new geographies of settlement in a context of transnational mobility (for Italy, see e.g. Lombardi-Diop 2009; Lucht 2011; Borri 2017), I am particularly interested in the ways in which, in the domain of high-risk migration, precarious migrant occupations become places where Black African migrants are actively trying to revitalize their attempts at making a life by re-establishing connections in a situation of increasing global disconnect.

So how is it that mobile subjects who have to navigate uncertain trajectories do or do not succeed in making meaningful places? While being part of a longer research trajectory, my argument in this chapter builds on a five-year multi-sited research in what Heather Merrill (2018), following Harney and Moten (2013), calls Black "fugitive sites"; or what, from an Africanist perspective, can be seen as the location of impermanent settlement: a togetherness in mobility that characterizes Black Africans' simultaneous dispossession and being in motion in a situation of ongoing rupture and uncertainty (Bolt 2015; Kleist and Thorsen 2017). While to some extent, I agree that such Black fugitive sites do represent a kind of "enraptured disclosure" (Harney and Moten 2013: 11, 28) from the logics of liberal governmentality, the desire of their inhabitants to be seen and recognized as legitimate political subjects also forces them to re-engage with the very forces that are responsible for their marginality through intricate relations of compromise, negotiation and concessions.

My empirical work for this chapter specifically emerges from a participant observation of one site, situated in Northern Puglia, called Casa

Sankara. Inaugurated as an autonomous *foyer*, or enclosure, by two Senegalese workers from the Gran Ghetto of Rignano Garganico, the site gradually became a central node in the latter's dismantlement as well as of the affirmation of a new form of 'neoliberal hybridity' in the form of externalized migration management. By following this process from the inside, I try to disentangle the intricate bureaucratic mechanisms and social interactions through which Black African segregation has continued to intensify in this fractured environment. But let me first tell a few words about the context in which this and similar occupations unfolded.

Bordering the Liquefied

My research question has to be contextualized within an open paradigm shift that has taken place in Mediterranean border management over the last half decade. In many cases, the active deregulation of asylum rights during this time has effectively reduced accountability towards migrants claiming asylum and access to human as well as territorial rights (Gammeltoft-Hansen 2013; Menz 2011, Kritzman-Amir 2011). Since an international coalition led by France and the UK staged a successful attempt to decapitate the regime of Muammar Gadhafi in the autumn of 2011, Libya has not only precipitated into protracted armed conflict, but its ports have also become principal departure points for Sub-Saharan African migrants trying to reach European shores. Together with the Tunisian Revolution, this event has generated what in Italy became known as the 'North Africa Emergency' (or *Emergenza Nord Africa*, ENA). For two years, migrants fleeing Northern Africa were automatically granted a temporary humanitarian residence permit on the peninsula. Persistent instability across the Central and Eastern Mediterranean routes (which included the Sicilian Channel and the Aegean Sea) caused European states—particularly those situated on the migration frontlines of this new war on migration, like Italy and Malta—to openly challenge the European border control system, for instance by assisting humans at risk and by granting asylum to those in need. After the closure of ENA, the subsequent *Mare Nostrum* operation, guided by Italian naval forces, saved at least 150,000 human lives (for a discussion about how this 'humanitarian operation' also conditioned a shift in the logic of assistance on firm land, see Heller and Pezzani 2016). Due to financial debts as well as growing pressures from European member states to tighten the grip over the Central Mediterranean migration route, *Mare Nostrum* was called to an end in October 2014 and replaced by Operation *Triton*, piloted by the European Border and Coast Guard Agency (or Frontex). This transition precipitated a rapid shift from saving lives to border security in Mediterranean migration policy.

In their active attempts to make the Mediterranean crossing drastically riskier, resulting in the proportional rise in migrant deaths (information available at https://missingmigrants.iom.int), individual EU member states have sought to forge active alliances with so-called transit states like Niger, Mauritania, Morocco, Egypt, Algeria and Tunisia to combat 'illegal migration' throughout this period. Existing agreements, for instance between Senegal, Mauritania and Spain, have been intensified, while new alliances, for instance between the late Renzi government and the incumbent Libyan government of General Haftar, have been established. This policy shift has not only involved an active transformation of humanitarian principles, but also a clear reconfiguration of the terms under which migration flows are regulated. From an active refoulement of migrants across Mediterranean waters, European member states are now practically outsourcing territorial border management to third states that are consequently becoming buffer zones for outward African migration. At the same time, these agreements also unambiguously make Europe complicit in migrant abuse. Amnesty International (2017) documented how hundreds of thousands of refugees and migrants trapped in Libya continued to be at the mercy of Libyan authorities, militias, armed groups and smugglers, who often work closely together for financial gain. Rather than a mere omission of the right to asylum, therefore, we are confronted now with a joint logistical effort to block, detain, refute and repress migrant mobility through methods of violent dispersal across the Mediterranean waters as well as on the African and European continents. (For a wider discussion, see Heller and Pezzani 2016.)

From 2011 to 2014, European governments firmly put in place a system of formal and less formal border controls that were meant to hamper migrant journeys on their way through the continent (Tazzioli 2018). The immediate consequence of these measures was a massive rise of migrant ghettos along main trajectories: from Calais to Ventimiglia, from Idomeni to Maljevac, migrants were stuck on their routes and put up temporary camps on Europe's territorial borders. During the 'Summer of Migration' in 2015, governments situated on the Balkan route—with the exception of Serbia—erected different kinds of barriers and border fences to block migrants on their journeys. In part, some EU member states also reacted against the deep political divisions that this stringent border regime was causing on the continent as a whole, with some governments, like Italy and Germany, actively welcoming refugees (and in some cases even omitting effective controls), and others, like Austria and France, refouling migrants at their borders. Altogether, therefore, it appears that the changing politics of asylum in Europe post-*Mare Nostrum* contributed to an active migrant dispersal, a politics which forced migrants across the EU-North Africa borderland to engage in ever more convoluted geographies of mobile navigation in order to challenge their constant illegalization and refoulement.

At the same time, the change of focus from saving lives to border surveillance has radically transformed the European border paradigm, from being mainly concerned with the channelling, filtering and categorization of cross-border flows on the continent towards a set of active strategies for considerably lengthening, diverting and deterring migrant journeys across multiple scales. But what has the consequence of this conscious dispersal of migrant flows been on existing patterns of migrant settlement in the southern European context? While by no means synonymous with the general condition of African immigrants in Europe, the fate of migrant travellers became increasingly intertwined with residing migrants who were already on the verge of clandestinity because of their irregular working conditions or lack of official residence papers. In that sense, the migrant ghetto has progressively taken the form of a relief valve (or a decompression chamber, to use Mezzadra and Neilson's terms: 2013: 149) for state governments, which neither have the will nor the policies in place to address such informal urbanity but at the same are happy to tolerate the existence of these sites as a way to both segregate and hide marginalized migrant subjects from public view—only to spectacularize their presence as signs of illegal trespassings and clandestinity. In this sense, territorial border infrastructures can be seen to create a "prolonged state of liminality" that ends up making the border indistinguishable from everyday life (Braverman 2011: 274).

It can be argued similarly in the European context how recent changes in asylum policy and border management practices are actively imposing a new infrastructural design that is effectively erecting a political and moral distinction between who is and who is not to be considered a righteous subject worthy of autonomy and (re-)habilitation (Balibar 2002). Indeed, one needs to ask how this distinction is not only made to work in everyday bureaucratic practice, but also how it simultaneously shapes and folds the embodied experience of migrants who have decided to lead their lives in Europe. This experience will be the centre of the following section, where

I detail my participant observation of one migrant 'ghetto' in San Severo, Puglia, over the course of 2015–18.

Casa Sankara

The first time I arrived in San Severo in 2015, Hervé came to meet me at the train station on a warm summer afternoon, his wide smile beaming towards me as I stood by the car. I introduced myself and my friend Marco, an architect and photographer who spent several years in Northern Africa. Previously, Hervé and I had spoken for a long time over the phone. He had welcomed me to visit Casa Sankara to meet (*palper du doit*) the project he and his uncle were trying to construe in the midst of many tensions and obstacles. Through a Skype video, I had briefly met the four families (in total fifteen people) who had found refuge there.

The site consisted of a few container buildings organized around a main square on the premises of a former agricultural enterprise in the outskirts of San Severo, on the road towards Foggia. In the background I could see kids running and families cooking their meals in what appeared like an African *foyer*. Upon arrival, Mbaye Ndiaye, whom Hervé later introduced to me as his uncle,[3] told me at length how they had moved to this place in 2013 when, after a fire destroyed part of the ghetto close to Rignano Garganico, they had found refuge in a local cooperative that was running an eco-village on site. Until very recently, the Gran Ghetto, as its name tells, was Italy's largest migrant ghetto.[4] Situated at the end of a potholed road far away from the tourist-trodden destinations of the Gargano coastline but in the heart of the agrarian region that gave rise to the first contentious struggles of Italian day labourers (led by Guiseppe Di Vittorio, a leader of Italy's most important labour union) in the early 1900s, Rignano has become one of the main Black fugitive sites in the region, along with Borgo Mezzanone (*la pista*), and Tre Titoli (*Ghana House*). The ghetto's historical origins are contested. According to the French anthropologist Benoit Hazard, who engaged in a participant observation of Burkinabe day labourers in the region in the early 2000s,

[3] Both were born into a family of *griots* from Senegal's west coast community of Lebous, close to Dakar. Both practice Sunni Islam through their adherence to two brotherhoods, the Mourid and the Tijani.

[4] In 2015, the *comune* of Rignano filed an official complaint over the name that continued to be used in the media, claiming (correctly) that the ghetto actually resides in the constituency of San Severo, more precisely in *contrada* Torretta-Antonacci.

the place was founded by a group of *bisa* workers from the southeast in 1996[5] after a fire destroyed the ghetto of Villa Literno two years earlier (Hazard 2004).

Along with other toponyms (like Boreano in Basilicata, Rosarno in Calabria, Castelvolturno in Campania), the *Gran Ghetto* constituted a central node in the circuit migrant labourers refer to as *le tube* or *le tuyau*. Besides serving as a stopover place and transit, the Gran Ghetto gradually assumed a more permanent function as a communal refuge in an environment of increasingly repressive migration laws and border controls. The ghetto could thus be described as a fluid form of urban dislocation: frequently destroyed, removed, and relocated. Both its architectural form and social components reappear over and over again in different places, but according to the same rules of conviviality, they altogether constitute a middle passage for the socialization of migrant travellers into the universe of clandestinity and invisibility of this new African diaspora (Carter 1997, 2010). Since its emergence, Rignano has grown exponentially into a veritable rural town, hosting several grocery shops, restaurants and brothels, and, in the summer, a radio station called Radio Ghetto in the commercial section of the ghetto, which has gathered the name New York. In contrast, the residential part, called Washington, hosted a series of neatly divided plots with plastic shelters constructed with recuperation material from Foggia's demolition sites. During the highest peak of the tomato harvesting season in August and September, when temperatures can rise up to 50 degrees Celsius, the ghetto counts up to 2500 inhabitants. Most have African origins and are facing extreme forms of exploitation as they engage in temporary jobs as agricultural day labourers. Besides the illegal businesses of the restaurant owners and labour gangmasters, the place also has attracted the attention of several non-governmental organizations over the years, from the medical charity Emergency to the labour unions USB and CGIL-FLAI and Catholic charities like Caritas and *Io Ci Sto* (I'm In)—an association run by a Swiss priest that does relief work.

Upon their arrival in the Gran Ghetto, Hervé and Mbaye were quite shocked to encounter this African *bidonville* in the heart of the Italian countryside, in the middle of Europe. In a video document they shot together with Francesco Belizzi (*La Belleville* 2015), they show me the

[5] The Bisa are a Burkinabe minority whose traditional kingdoms are located close to the Ghanaian border in the region of Boulgou (Hazard 2004).

miserable conditions of the place. Their friend Nadine, who stays with them in San Severo, expresses the same perplexity. Being brought to Rignano under false premises, she says she felt depressed by the misery and exploitation of her African brothers and sisters, and she was forced to sell her body in the ghetto's restaurants.

Omar, who joins us later that evening, tells me about the mud that conditioned his every move in the ghetto and how he sat for hours in his shack, depressed, thinking of how to get out of this place. He remembers with nostalgia his time in Libya, where he was able to work every day at construction sites and send money home. He did not have to care about the cost of electricity, a car or accommodation, because everything was paid by the government. When the war came, he did not want to leave, but he was finally forced on a boat by the military. For two years they let him stay in a reception centre in Campania *a mangiara pasta al pomodoro*—and wait for his documents. With the little pocket money he received from the cooperative, he could go to the local supermarket but the residents resisted selling him anything. He is angry with the Italians and the French for having blown up Libya and created this mess: "Now they lament themselves that too many migrants are arriving at their doorsteps. But who has created this problem, this chaos in the first place? It is them!" he concludes.

The experience of Mbaye, Hervé, Nadine, Omar and the other inhabitants of Casa Sankara in San Severo is representative of the challenges the new African diaspora faces in Italy in the aftermath of the Libyan refugee crisis. For a period of eight months in 2013–14, they were engaged by an Italian cooperative to host migrants who kept arriving in the context of the *Mare Nostrum* operation; ninety of them lived permanently on the Casa Sankara site. In practice, they became the cooperative's employees, taking care of the accommodation, food and relations to public institutions of "those people who live the conditions of marginalization on their proper skin" (Skype interview 25 March 2015). But in practice, they frequently repeat during our meetings, "we were confronted with a mentality of exploiters" (Lamin, 28 July 2015). In the context of the growing privatization of migration management, the Italian government started to assign more structural funds to all sorts of cooperatives, voluntary groups and associations, in order to spread the growing population of asylum seekers and refugees across its territory. Through public bidding processes that were not always completely transparent, these funds also became a way to buy the loyalty of political clients and distribute economic benefits across patronage networks. The result was a rapidly spreading clientelistic network of 'red' and 'white' cooperatives that casted their wings over this emerging sector of migrant reception, a negotiation that often involved the highest levels of regional and national governments.[6] Because of the political backing he received at the regional government, the president of the cooperative apparently kept using the resources he received for the reception centre as if it were a personal fund, pretending his employees were voluntary workers, and thus denying them a contract. This eventually led to an open conflict. Hervé and the other African members protested they desperately needed the contract to renew their residence papers, to maintain access to social services, but also generally to retain their legal status in the country. They accused the president of using Casa Sankara, which they had built, for his personal benefit.

The conflict between the cooperative and its African workers quickly led to a war of position between the president and his other Italian cooperative members, the representative of the labour union CGIL, and some former members of the town administration on the one hand; and the diaspora's support network on the other. The former include the hitherto prefects of Foggia, Luisa Latella,[7] the lawyers' collective *Avvocati di Strada*,[8] the town counsellor for legal affairs and future regional governor Michele Emiliano, and his close collaborator Stefano Fumarulo. Importantly, the inhabitants of Casa Sankara could also count on the help of local police officers, whom—despite the fact that many employees had their residence permits expired since 2014, intervened in their favour on crucial moments. In another video document Hervé shared with me, the chief constable of police actively blocked the attempt of the cooperative's president to evict Casa Sankara's inhabitants from their container homes with the assistance of two plumbers. As a result, a kind of informal modus

[6] The most publicized case in this domain was no doubt the *mafia capitale* investigation in Rome (for a discussion see Abbate and Lillo 2015). In my interview with the regional administration in Puglia, a close aid of the governor denounced the 'parallel system' of corruption that characterized migrant reception: "Puglia is a territory colonized by associations that work for the most part at the expense of the immigrant and make them into their own business" (26 August 2016). Various press reports during this period continued to refer to corrupted bidding processes and cooperatives delivering suboptimal services.

[7] Interview, 16 August 2016. Significantly, Latella also assisted the mayor of Riace (Calabria) Domenico Lucano, who became famous for his refugee protection programme during the same period.

[8] During 2015 and 2016 I had several interviews with AdS representatives in Bologna and Cerignola, including with Casa Sankara's former lawyer Stefano Campese.

vivendi ensued, which was based on the mutual social recognition that, though things were not entirely legal in this place, a certain legitimacy characterized the activities going on in this diasporic urbanity, including its very ownership by less-than-legal subjects.

Overall, therefore, one could argue that the kind of "invisible governance" (Hecht and Simone 1994) of this fugitive site reflected the extremely ambivalent condition under which its inhabitants had to negotiate their lives between agents who actively undermined the authority of public institutions they reaped for their personal benefit and marginalized subjects who, on the contrary, depended on official legitimation to claim access to state protection. At a deeper, psychological level, Casa Sankara's inhabitants found themselves in a situation where they were faced with the simultaneous need to re-build relations from scratch in a world that was lacking orderly state regulation. But they also desperately needed to relate their efforts to the "commanding heights" (Hart 2009: 27) of political power in the corrupt fortunes of public office-holders who not only owned the means to protect them, but also directed their access to a direly needed political status and recognition. In this legal and psychological limbo, they were nonetheless capable of challenging the pretended cultural and political superiority that the cooperative's president and members of the local political elite pretended to have over them, and in their resistance even attracted other important activists who partly contested this state of affairs in the name of legality. Hervé and Mbaye had two solid reasons for doing so.

Becoming European

First, the two Senegalese activists made it plain that they were not going to surrender to the stereotypical ascription that dominated the discourse of Italian human rights defenders, and which depicted African migrants simply as victims in need of assistance: "We were not people in need who he [the president of the cooperative] found in the street like that," Hervé said. "We live in San Severo, this is our home" (Interview 27 July 2015). This explicit reference to place contained both a very political and unconventional statement in the landscape of the Mediterranean 'migrant emergency'. In a context of apparent fluidity and volatility, their new rootedness in Casa Sankara constituted a crucial claim to territory and belonging.

Repeatedly pointing out the difference between *migrante* (migrant) and *immigrato* (immigrant), I asked them whether they ever lost the gusto to continue their struggle to find a place. Responding affirmatively, Mbaye recalled how once he received a retribution from a court case he had won back in Senegal: "You know, I am a famous man there," he said, "I own a business, and I used to be married to the most famous pop singer of the country." But the purpose they had set themselves obliged them to stay put. Hervé added, "We say Casa Sankara already has our tombs... we won't leave anything behind. [*Noi diciamo, a Casa sankara ci sono gia' le nostre tombe... Non lasciamo niente.*]" (Interview 27 July 2015). Because he had been fasting for over three months to observe a strict religious diet, Mbaye only occasionally exited his room in the container. Exhibiting his religious dedication, he told me, "We have to be on the side of legality. [*Noi dobbiamo stare nelle legalità.*]" (Interview 29 July 2015).

This declared insistence on legality opened the door to a new logic of the struggle Mbaye and his companions were engaged. On the one hand, they continued to challenge the convention that had been signed between the local government and the cooperative, and which had resulted in their physical exploitation. On the other hand, they also denounced the criminal gangmaster rings that continued to be the kingpin of the local agrarian economy in Northern Puglia, and against whom Di Vittorio had initiated his revolution in the early 1900s. At the risk of their own lives, they agitated against the African *capineri* (literally: Black bosses) who organized informal labour and illegal prostitution in the Gran Ghetto. Notwithstanding this strong opposition, they were involved in tense mediations between the criminal intermediaries who governed the ghetto's illegal businesses and those people who actively wanted to get out—including one *maman* (brothel owner) who ended up living in Casa Sankara with her husband and seven-year-old son. These actions gradually gained the support of the regional government and judiciary, who waged an active campaign against Rignano through a joint programme called *Ghetto Out – Capo Off*, but which had difficulty in getting off the ground amidst numerous sources of opposition.

Second, Casa Sankara's plight involved a strong reference to transnational belonging. In contrast to the criminal politics and entrenched social relations they observed all around them, the insistence on being mobile and rooted in different places became one of force rather than weakness. "We do not have to take lessons in democracy from anyone," Mbaye said to me in a subsequent conversation. "We are born democrats. [*Siamo nati* 'démocratique']" This statement, which is also interesting from a sociolinguistic viewpoint (in fact, Mbaye kept on switching between his native tongue, French, and Italian) highlights a vivid adherence to a diaspora politics—call it a form of black reason (Mbembe 2017)—which simultaneously agitates against the Black man as an objectified, extractive body, and enlightens the permanent shadows in which it is being casted. Strikingly, Mbaye, Hervé and the other African diasporic leaders who emerged in this context started to connect their discourse to a wider *Afrotopian* ideology, which, in the words of its inspirator Felwine Sarr, vehemently charges against the stereotypical representation of Africa as a reservoir of misery and of deficiency, instead underlining the necessity to reformulate an African politics based on its own terms. Altogether, these strands appear to be arguing in favour of an "epistemic decentring" (Sarr 2016: 107) which, in contrast to the "being-like" in the world that continues to be imposed through Eurocentric development indexes and modes of government, delineates a different teleology based on a different imaginary.

Indeed, for some time it appeared that Casa Sankara was actively building its own utopia, giving itself the task of discovering a potential for transformation in this condition of diasporic fugitivity. Not unsurprisingly, this quest gave considerable weight to some of the cultural and, especially, spiritual values that had been deemed to be lost but were now actively reinvigorated as the "the anchor of the dynamic of the human becoming of the African man" (Sarr 2016: 32). An important aspect of this reinvention involved the promotion of Mourid culture, which, in the context of the Senegalese diaspora, has already been described as a central unifying factor (Riccio and Degli Ulberti 2013). By making Casa Sankara a central node in the organization of religious gatherings and cultural events, its Senegalese inhabitants were actively building an archive that simultaneously sought to provide a counter-narrative to the myth of unique, hegemonic modernity, while also "bringing to matter" what has been denied existence (Mbembe 2017: 47).

Hervé remembered an important event in this regard, which literally placed him and his fellow activists at the heart of the European struggle for labour rights. During a grand gathering of the labour union FLAI-CGIL in Foggia, where the issue of migrant exploitation first came to the fore, he was called upon to let his voice be heard. In a room packed with political representatives, a local union representative pushed Hervé to the stage, telling him that he had one minute to make his point. Hervé said: "Behind your food lays our slavery. [*Dietro il vostro cibo c'e' la nostra schiavitu'*]." With this statement, Hervé effectively filled the ranks of a wider struggle that had started some years earlier under the leadership of other African workers. One of these was Yvan Sagnet, author of two books and a former engineer student, who had occupied a former *masseria* (farm) in Puglia with other African farmworkers in 2011. For months, they denounced their exploitation by Italian and African *caporali*: ruthless gangmasters who thrived on the former's dependency on informal labour in Italy's food agriculture. Since then, regular riots had erupted across the country, directed both against and in favour migrant workers' rights. In Rome, the Ivorian activist Aboubakr Soumahouro similarly denounced the abusive working and living conditions of Italy's migrant reception centres. Together with other foreign and Italian activists, he co-founded the Baobab Experience, which, similarly to Casa Sankara, was trying to raise an autonomous constituency of migrant resistance in the face the closing border regimes in Italy and broader Europe. The protagonists of these struggles kept invoking revolutionary figures like Thomas Sankara and Guiseppe Di Vittorio to provide a symbolic bridge between African and European political cultures. In Hervés words, "the immigrant has to be an interlocutor, protagonist of his own fate" ["*L'immigrant doit etre un interlocuteur, protagoniste de son propre sort*"] (25 March 2015). In his plea, however, African political liberation could not take place without actively constructing a bridge with European interlocutors.[9]

And here lay exactly the ambivalence of their claims. The need to insert its discourse into a locally embedded politics of legality meant that Casa Sankara effectively took over the frame of Puglia's official institutions—particularly the new regional governor and procurator, who continued to insist on the criminal aspects of the ghetto as a free zone (*zona franca*) placed outside civilization (*fuori della civiltá*).[10] This attitude gradually distanced them from other arguably more radical groups, whose protagonists continued to insist on the wider scales of African exploitation and the

[9] Coincidentally or not, this need to negotiate African identity in Europe had been exactly the position of the Senegalese writer and co-founder of the diaspora journal *Présence Africaine*, Alioune Diop, whom during the 1959 Congress of Negro Writers and Artists in Rome pointed out that "we can bring out the features of our personality only through a dialogue with the West." Gilroy cites this passage as a key moment for the emergence of an Afro-European consciousness in his book *The Black Atlantic* (1983: 196).

[10] Interview with Stefano Fumarulo, 26 August 2016; Interviews with Hervé 15 January 2016, 28 August 2016.

need to resist institutional racism. One of these was the aforementioned Aboubakr Soumahouro, who in one public meeting I attended late August 2016 emphasized the deliberate exploitation of social despair that multinational enterprises continued to thrive on in the agricultural industry. Reminiscent of Loic Wacquant's proposition about the American ghetto,[11] Soumahouro remembered how Africans are being placed in an oppressive triangle of spatial segregation, legal marginalization and social repression. He and others resisted a simplistic legal reading of the phenomenon of labour exploitation by insisting on the monetary and political gain that was to be made from oppressing, not just African, but *all* irregular workers. In his public interventions, Yvan Sagnet, in contrast, highlighted the important legal battle that needed to be waged to combat illegal hiring practices and emphasized the need to work with state institutions towards thorough legal reform.

Given these multiple battles, it is counterproductive to locate the implications of Casa Sankara's struggle completely outside a much wider conflict African migrants were involved in in Italy at the time and which also concerned the reconfiguration of Italian politics in the face of the current crisis of liberal democracy. As already mentioned, Casa Sankara did not constitute the only alternative to the African ghetto, as Mbaye and Hervé systematically upheld. But instead it was located in an archipelago of smaller sites and constituencies whose predicaments depended to a great extent on their active representation by others. In the neighbourhood of Foggia alone, another dozen ghettos were fighting for visibility. Among these was Casa Bianca, a former milk factory occupied by a loose community of African youngsters with the assistance of the Italian association Pro Fuga. In the building, a group of around seventy young adults divided joint tasks of cooking, gathering firewood, and looking for day jobs in the neighbourhood, while insisting on their autonomy from more

[11] Wacquant (1997: 342) famously described the American ghetto as a "frontier territory"; more concretely as "… a historical determinate, spatially-based concatenation of mechanisms of ethnoracial closure and control." His key point is that one cannot simply "read off" the ghetto's social structure from its supposed deficiency and lack of formal institutions, but one must positively identify the principles that underlie its internal order and govern its specific mode of functioning. The analyst must explicate and display in some detail the concrete mode of structuring social relations and representations operative within the ghetto—the work of collective self-reproduction—whereby its residents endow their world with form, meaning and purpose, rather than simply report that this mode differs from those that hold sway in other sectors of society.

institutionalized migrant reception centres.[12] In other ghettos, which persisted more distantly form the public eye in Foggia's rural hinterlands, migrants had to constantly navigate their access to social services between different, and often opposing, institutional networks and their repertoires, including Italian voluntary organizations, ethnic networks and gangmasters. At the end of the day, therefore, this spatial segregation not only reproduced the convoluted geographies Black African migrants were forced to engage in in order to be seen by the state. But it also stimulated a rising tensions over who effectively *owned* the ghetto and its exploited inhabitants.

Between the more or less institutionalist perspective of Casa Sankara and other groups, and the 'autonomous' perspective of radical Italian associations, a war of position unfolded across multiple scales, which gradually led to a breaking point. The open conflict that African and Italian protagonists were engaged in to represent the interests of the ghetto eventually also increased the already existing mistrust towards regional state institutions, which were progressively depicted as trying to lure the remaining *ghetissards* into submission. An important moment in this regard constituted a protest organized in Foggia early 2016, where migrant workers openly contested their eviction from the ghetto. One of them held a billboard saying "The ghetto is our home [*Il ghetto e' casa nostra*]." The statement was emblematic not just of the segregation African migrant workers were being subjected to in this context. It also showed how the government's metaphorical equation of the ghetto with mere criminality was bumping into increasing resistance not just from the *caporali*, but from a wider constituency of African migrants. In the meantime, the proposition of Casa Sankara to construct an independent African cooperative was increasingly being shattered in favour of its proceeding reconversion into a migrant reception centre, which, under a thin veneer of African autonomy, reproduced exactly the same conditions that its protagonists had so vehemently resisted. A number of clear political reasons can be cited for things to unfold in this way, which I will explain in the next section before reaching some general conclusions.

[12] Individual interviews and focus group discussion on site on 2 and 3 July and 29 August, 2016.

Conclusion: Concessionary Politics

In January 2017, the regional administration in Puglia finally decided to materialize its decision to evict Rignano for good. During the eviction, a massive fire—probably instigated by arsonist *caporali*—destroyed much of its remains before bulldozers tore down the remnant dwellings. Accompanied by the winter rain, inhabitants were transported in small groups to the premises of Casa Sankara, which in the meantime had signed an official convention with the regional government as a temporary migrant host centre. (In the convention, the site was renamed Azienda Fortore, its original name, to avoid confusion with the previous cooperative.) This move was the last stage of preparations that had taken more or less a year's time.[13] After two years of legal limbo, this moment constituted a turning point: We are envisaging a concrete change now ["*On vise vers le concret maintenant*,"] Hervé told me (13 April 2016), adding that the imminent hosting of over 150 evicted *ghetissards* on its premises obviously required a different, managerial logic: "Things have to be organized more formally now," he told me (6 June 2016). By this, he meant essentially that the institutionally backed access to the premises and its services of free shelter, food, and social services for migrant hosts were a first step towards the realization of their objectives of autonomy and self-sufficiency. But more and more, doubts arose whether this proclaimed autonomy had not made place to an indirect form of externalized migration management. While the Italian civil protection service came by to deliver their standard field beds to the tents and the top floors of the building, the ground floor was quickly renovated to host a cafeteria, a kitchen, and a storage room. Once the new inhabitants arrived, the gate of Casa Sankara remained firmly closed for a couple of

[13] For almost a year, the site of the Grand Ghetto had been placed under judicial investigation. After a temporary eviction, in which over 300 *ghetissards* were arrested and brought to police headquarters (but only a minority of 30 of them further detained for having no valid papers), the judged ordered what he called a *perquisizione con modalita' d'uso*. While some arrests (including one of an Italian drug kingpin) were indeed made in the context of this ongoing investigation, in practical terms, illegal activities—including prostitution and illegal labour hiring—could continue unabated. Interestingly, the procurator of Foggia publicly distributed evidence for the instigation of the fire before the judicial investigation was concluded. This included a video fragment were a Senegalese man cynically comments the plight of two youngsters who were killed in the fire. (I have analysed the fragment on my blog www.timothyraeymaekers.net). The two youngsters, Mamadou Konate and Nouhou Doumbia, were widely mourned across African communities in Italy as a sign of solidarity.

weeks, only to be opened upon direct invitation by the door managers. ("We have to make sure we keep out the *caporali*," Hervé told me). Each time I came back to visit in 2017 and 2018, I saw more and more tents appearing in Casa Sankara's former grocery garden, which Lamin and others had hitherto proudly shown to me as the foundation of their new cooperative. Hervé, Mbaye and the other protagonists of Casa Sankara were seen on site less and less. The 35 hectares of Azienda Fortore, which Mbaye had initially intended as an experimental site for his agricultural cooperative, were now professionally managed by a local firm. Besides the usual African hairdresser and grocery shop, there was in fact very little one could see develop in the form of an African community at Casa Sankara. Instead the place looked more and more like a migrant-run reception centre, which—despite its nominal independence—fitted exactly into the model of a privatized, hybrid but at the same time decisively neoliberal institution in a context of rising anti-immigrant protests and right-wing populism.[14] In the meantime, it was evident that the ghetto of San Severo was not going to disappear overnight. With the new agricultural season approaching, workers were already inventing alternative shelters and sites, including mobile homes and caravans, which gradually occupied former ghetto plots. Instead of eradicating the ghetto, therefore, its eviction had just led to a further dispersion of African migrants on European soil.

Rather than resolving the migration crisis, the process of eradicating its informal migrant settlements thus brought Puglia's policy in line with the active politics of migrant diffusion that EU countries, including Italy, were implementing across the Mediterranean channel. The actual dispersal of migrant subjects in this context constituted a logistical element of an apparently concerted effort to negate and suppress migrant rights in this rapid reconfiguration of Europe's border surveillance regime. While some would refer to this process as a form of neoliberalization (see Gammeltoft-Hansen 2013; Kritzman-Amir 2011), the more original question to ask in this context, however, is: what has this active withdrawal of state institutions from governing migrant asylum produced in terms of the latter's

[14] During one interview, the regional administrator responsible for Casa Sankara explained to me how the idea about self-reliance had arisen: during a personal visit to Latin America, he became acquainted with the *para migrantes* hostels, where hosts actively assist in the maintenance and daily management of these sites (Interview 10 February 2016).

right to life? In the case of the eviction of the Gran Ghetto and its 'alternative' model of Casa Sankara, I argue that a more intricate process of an active renegotiation and contestation of this right has been going on, which makes it difficult to sustain a thesis of outright delegation of authority to arguably 'privatized' institutions.

To some extent, the case of Italy's migrant reception regime appears to reflect the transition from a governmental state, with its focus on calculation, surveillance and active regulation, towards a society of control, which, rather than strict enclosures, engages in the attempted modulation of processes that are already underway, a series of punctual interventions in a continuously changing network of associations (Deleuze 1992; Amoore 2013). Given the highly politicized character of these modulations, however, a more closely applicable terminology in my view is that is that of a government concession. Contrary to enclosures or *foyers*, concessions can be regarded as "indiscrete sites" which explicitly "thicken" politics around the arbitration of regulatory force (Cote and Korf 2018: 466). In contrast to more strictly regulated borders and checkpoints, they intensify the political struggle over who gets to decide the terms: not just of the force of law, but also of their concrete implementation in a bounded, territorial space. The terminology of concessions can help us understand how, in a context of legal pluralism where governance processes are disputed between contrasting institutional and cultural repertoires, the debate on who gets to *own* this process of rewriting the space of asylum and refuge is not predetermined. In the Italian context, migrant ghettos give rise to serious political frictions, which typically results in open contest, disagreement, and confrontation between various stakeholders on different scales. The often unexpected and unpredictable manner in which this negotiation unfolds has been the main interest of this paper.

The case I discussed here suggests that territorial borders—with the Mediterranean border being a case in point—do not constitute anomic or empty spaces, where the division between the included and excluded is written into the exclusive territorial order of the nation-state. Nor are they mere lines in the sand, which merely permeate the indistinction of rightless subjects in the face of citizens who can express their political rights. Instead they constitute vibrant spaces arousing with life and the battle for it between often contrasting agencies and institutions. Through their temporal and geographic expansion, territorial borders indeed may become extensive liminal spaces, where the right to belong and be recognized as a political subject is actively negotiated, contested, and submitted to compromise. As it goes in such places, the regulations and repertoires that govern 'normal' life may be—and, in the current Mediterranean context *are* indeed—seriously disrupted. This process simultaneously produces a widening margin for alternative regulations to arise in the midst of violent renegotiations, rising tensions and contestations. It is exactly in this context that the emergence of a new Black Mediterranean subjectivity needs to be situated. As I wrote, this subjectivity is partly inspired by a new, Afrotopian ideology that locates the history of African Blackness again in a firm, place-based setting and imaginary. The will to be seen and recognized politically, however, is forcing the Black Mediterranean into a simultaneous compromise of hybrid governance and political identification of which the outcomes are yet to unfold. What is certain, though, is that these will be carried both by the desires to narrate a counter-discourse and out of resistive belonging and identifications, as by important connections and disconnections, cultural crossovers and contaminations (Chambers 2008: 25; Giaccaria and Minca 2011; Lahoud 2013).

In the case I discussed here, two processes can be observed at the same time. On the one hand, a conscious process of eradicating an existing concession has been pushed through. Over the past five years, certain clientelistic networks in the domain of migrant reception, as well as the various extractive economies that had developed around it, have been deliberately criminalized and depicted as 'illegal' and 'anti-social' (or worse, void of civilization). On the other hand, the will to modify migration governance from the part of several stakeholders has motivated government authorities to grant an alternative, formalized concession to an arguably more obedient network of organizations, while nonetheless refusing to take full responsibility over the process of pacifying this contentious 'frontier'. In trying to capture this space, all stakeholders in fact have been involved to some extent in the 'making of' concessions, which typically comprised a great deal of compromise, negotiation and dilution between those subjects who are deliberately placed outside the metropolitan centres of liberal society, but who somehow live in the hope to enable ruling elites to gain from this process by formalizing and legitimizing their claims. Altogether, however, the deliberate decision of Casa Sankara to underwrite this dynamic ended up reproducing a peculiar form of compromise that reinforced, rather than resolved, the indeterminate status Black African people living on the European shores of the Mediterranean continue to endure. This persistent ambivalence, I argue, is a central feature of a changing paradigm of bordering the Mediterranean in the context of its proceeding privatization and outsourcing to increasingly unaccountable bodies.

Over the last half decade, we have been observing a rapid transformation of bordering infrastructures across the Mediterranean that has not just involved a radical renegotiation and denigration of human rights, but also a reconfiguration of modes of life. Black lives constitute not the only, but definitely a predominant, node of transformation in this context. In different ways, the so-called 'refugee crisis' that followed the outbreak of armed conflict in Libya has abruptly thrown southern Europe into the postcolonial age. The paradigm of the Black Mediterranean allows us to place this dynamic in perspective: not just of changing geopolitical relations across the Middle Seas, but also for explaining how this liquid border is transforming society beyond its territorial infrastructure. While the various attempts to un-govern migration management and the right to asylum have certainly produced a high degree of social erasure, these have also been met with a real attempt at re-territorializing identities in a context where migrants', specifically Black Africans', political subjectivity is systematically denied access to the arguably 'modern' spaces of history, politics and culture (Hanchard 1999; Carter 2010). As I have shown, the impermanent sites that are conceded to temporarily erect their tents under the constant threat of eviction also constitute a partial attempt to create a sense of being connected, or rather to *re*-connect, in a context of deep social and geographic rupture. More concretely, the story of Casa Sankara and the eviction of the Gran Ghetto I described in this chapter could be regarded as one of several interconnected nodes, or passage points, from where contemporary Mediterranean social histories are being actively rewritten. Rather than constituting an anomic or an empty space, the concessions that emerge in this confluence of events will obviously continue to generate lots of tensions between Italian and African aspirations and desires, and between different perspectives on who controls the right to decide over migrants' destinies. Altogether, therefore, these nodes present us with fundamental questions about the kinds of infrastructures that sustain human mobility in the context of reconfiguring border regimes, as well as the forms of community that emerge in their midst.

References

Abbate, L., and M. Lillo. 2015. *I re di Roma: destra e sinistra agli ordini di mafia capitale*. Roma: Chiaralettere.

Agier, M., and C. Lecadet. 2014. *Un Monde de camps*. Paris: La Découverte.

Amnesty International. 2017. Libya's Dark Web of Collusion: Abuse Against Europe-Bound Refugees and Migrants, AI (Index Number: MDE 19/7561/2017).

Amoore, L. 2013. *The Politics of Possibility: Risk and Security Beyond Probability*. Durham: Duke University Press.

Balibar, E. 2002. What Is a Border? In *Idem, Politics and the Other Scene*, 75–86. London: Verso.

Bolt, M. 2015. *Zimbabwe's Migrants and South Africa's Border Farms: The Roots of Impermanence*. Johannesburg: Wits University Press.

Borri, G. 2017. Humanitarian Protraction Status. The Production of (Im)Mobile Subjects Between Turin and Berlin. In *ERQ*, gennaio–aprile, pp. 55–73.

Brambilla, C. 2014. Shifting Italy/Libya Borderscapes at the Interface of EU/Africa Borderland: A "Genealogical" Outlook from the Colonial Era to Post-Colonial Scenarios. In *The Mediterranean Migration Frontier*, ed. T. Raeymaekers, in: *ACME: An International E-Journal for Critical Geographies* 13 (2): Special Issue.

Braverman, I. 2011. Civilized Borders: A Study of Israel's New Crossing Administration. *Antipode* 43 (2): 264–295.

Bredeloup, S., and M. Zongo. 2005. Quand les frères burkinabé de la petite Jamahiriyya s'arretent à Tripoli. *Autrepart* 4/36: 123–147.

Brenner, N., and N. Katsikis. 2013. Is the Mediterranean Urban? In *Implosions/Explosions: Towards a Study of Planetary Urbanization*, ed. N. Brenner, 428–459. Berlin: Jovis.

Carter, D.M. 1997. *States of Grace: Senegalese in Italy and the New European Immigration*. Indianapolis: University of Minnesota Press.

———. 2010. *Navigating the African Diaspora: The Anthropology of Invisibility*. Minneapolis: University of Minnesota Press.

Chambers, I. 2008. *Mediterranean Crossings: The Politics of Interrupted Modernity*. Durham: Duke University Press.

Corrado, A., C. Castros de Carranza, and D. Perrotta, eds. 2016. *Migration and Agriculture: Mobility and Change in the Mediterranean Area*, Routledge ISS Studies in Rural Livelihoods. London: Routledge.

Cote, M., and B. Korf. 2018. Making Concessions: Extractive Enclaves, Entangled Capitalism and Regulative Pluralism at the Gold Mining Frontier in Burkina Faso. *World Development* 101: 466–476.

Davies, T., S. Dhesi, and A. Isakjee. 2017 forthcoming. The Rise of the Informal Refugee Camp: New Exclusions in Europe. In *Handbook on Critical Geographies of Migration*, ed. K. Mitchell, R. Jones, and J. Fluri. Cheltenham: Elgar Publishing.

Deleuze, G. 1992. Postscript on the Societies of Control. *October* 59 (Winter): 3–7.

Fontanari, E. 2017. It's My Life. The Temporalities of Refugees and Asylum-Seekers Within the European Border Regime. In *ERQ*, gennaio-aprile, pp. 25–54.

Gammeltoft-Hansen, T. 2013. The Privatisation of Migration Control and State Responsibility. In *Access to Asylum: International Refugee Law and the Globalisation of Migration Control*, 158–208. Cambridge: University Press.

Giaccara, P., and C. Minca. 2010. The Mediterranean Alternative. *Progress in Human Geography* 35 (3): 345–365.

Gilroy, P. 1983. *The Black Atlantic: Modernity and Double Consciousness*, Harvard: Harvard University press.

Giordano, C. 2008. Practices of Translation and the Making of Migrant Subjectivities in Contemporary Italy. *American Ethnologist* 35/4: 588–606.

Hanchard, M. 1999. Afro-Modernity: Temporality, Politics, and the African Diaspora. *Public Culture* 11 (1): 245–268.

Harney, S., and F. Moten. 2013. *The Undercommons: Fugitive Planning & Black Study*. New York: Autonomedia.

Hart, K. 2009. *Between Bureaucracy and the People: A Political History of Informality*. Danish Institute for International Studies, Working Paper, 2008/27. Copenhagen: DIIS.

Hawthorne, C. 2017. In Search of Black Italia. *Transition* 123 (2017): 152–174.

Hazard, B. 2004. Entre le pays et l'outre-pays. *Journal des africanistes* 74 (1/2): 2–15.

Hecht, D., and A.M. Simone. 1994. *Invisible Governance. The Art of African Micro-Politics*. New York: Autonomedia.

Heller, C., and L. Pezzani. 2016. Ebbing and Flowing: The EU's Shifting Practices of (Non-) Assistance and Bordering in a Time of Crisis. In Near Futures Online 1: Europe at a Crossroads.

Kasparek, B. 2016. Complementing Schengen: The Dublin System and the European Border and Migration Regime. In *Migration Policy and Practice: Interventions and Solutions*, ed. H. Bauder and C. Matheis, 59–78. New York: Palgrave.

Kleist, N., and D. Thorsen, eds. 2017. *Hope and Uncertainty in Contemporary African Migration*. London/New York: Routledge.

Korf, B. and T. Raeymaekers. 2013. Border, frontier, and the geography of rule at the margins of the state. In *Violence on the margins: State, Conflict, and Borderlands*, ed. B. Korf and T. Raeymaekers. New York: Macmillan.

Kritzman-Amir, T. 2011. Privatization and Delegation of State Authority in Asylum Systems. *Law & Ethics of Human Rights. Private Power and Human Rights* 5 (6): 203.

Lahoud, A. 2013. The Mediterranean: A New Imaginary. In *New Geographies, 5: The Mediterranean*, ed. A. Petrov, 81–100. Cambridge, MA: Harvard University Press.

Lecadet, C. 2013. From Migrant Destitution to Self- Organization into Transitory National Communities: The Revival of Citizenship in Post-Deportation Experiences in Mali. In *The Social, Political and Historical Contours of Deportation*, ed. B. Anderson et al. New York: Springer.

———. 2014. Ghettos: Deportation Without a Voice. In *Migration: The COMPASS Anthology*, ed. B. Anderson and M. Keith, 157–159. Oxford: COMPASS.

Lombardi-Diop, C. 2009. Roma Residence. *Interventions* 11 (3): 400–419.

Lombardi-Diop, C., and C. Romeo. 2015. Italy's Postcolonial 'Question': Views from the Southern Frontier of Europe. *Postcolonial Studies* 18 (4): 367–383.

Lucht, H. 2011. *Darkness Before Daybreak African Migrants Living on the Margins in Southern Italy Today*. Berkeley: University of California Press.

———. 2017. Death of a Gin Salesman: Hope and Despair Among Ghanaian Migrants and Deportees Stranded in Niger. In *Hope and Uncertainty in Contemporary African Migration*, ed. N. Kleist and D. Thorsen. London/New York: Routledge.

Mbembe, A. 2017. *Critique of Black Reason*. Trans. Lauren Dubois. Durham: Duke University Press.

Menz, G. 2011. Neo-liberalism, Privatization and the Outsourcing of Migration Management: A Five-Country Comparison. *Competition and Change* 15 (2): 116–135.

Merrill, H. 2018. *Black Spaces: African Diaspora in Italy*. Oxan/New York: Routledge.

Mezzadra, S. and B. Neilson. 2013. *Border as method, or, the multiplication of labor*. Durham and London: Duke University Press.

Papadopoulos, D., N. Stephenson, and V. Tsianos. 2008. *Escape Routes. Control and Subversion in the 21st Century*. London: Pluto Press.

Pinelli, B. 2017. Borders, Politics and Subjects: Introductory Notes on Refugee Research in Europe. In *ERQ*, gennaio-aprile, pp. 5–24.

Raeymaekers, T., ed. 2014. The Mediterranean Migration Frontier. *ACME: An International E-Journal for Critical Geographies* 13 (2): Special Issue.

Riccio, B., and Uberti S. Degli. 2013. Senegalese Migrants in Italy: Beyond the Assimilation/Transnationalism Divide. *Urban Anthropology and Studies of Cultural Systems and World Economic Development* 42: 207–254.

Roy, A. 2005. Urban Informality: Toward an Epistemology of Planning. *Journal of the American Planning Association* 71 (2): 147–158.

Sarr, F. 2016. *Afrotopia*. Paris: Philippe Rey.
Saucier, K., and T.P. Woods. 2014. Ex Aqua. The Mediterranean Basin, Africans on the Move and the Politics of Policing. *Theoria* 141/61/4 (December): 55–75.
Tazzioli, M. 2018. Containment Through Mobility: Migrants' Spatial Disobediences and the Reshaping of Control Through the Hotspot System. *Journal of Ethnic and Migration Studies* 44 (16): 2764–2779.
Thomassen, B. 2014. *Liminality and the Modern: Living Through the In-Between*. London/New York: Routledge.
Ticktin, M. 2016. Calais: Containment Politics in the 'Jungle'. *The Funambulist: Politics of Space and Bodies* 05 (May–June): 28–33.
Wacquant, L. 1997. Three Pernicious Premises in the Study of the American Ghetto. *Events and Debates*, pp. 341–353.
Yiftachel, O. 2009. Critical Theory and 'Gray Space': Mobilization of the Colonized. *City* 13 (2–3): 240–256.

CHAPTER 8

"These Walls Must Fall": The Black Mediterranean and the Politics of Abolition

Ida Danewid

> If their blood has not mingled extensively with yours, their labour power has long since entered your economic blood stream. It is the sugar you stir, it is in the sinews of the infamous British sweet tooth, it is the tea leaves at the bottom of the British cuppa. (Stuart Hall)[1]

Introduction

Calls for closed borders are increasingly heard from all sides of the political spectrum. Where restrictive immigration policies once were the sole prerogative of the Farages, Le Pens, and Orbáns, they now find widespread support on the social democratic left. As Hillary Clinton explained during a trip to Europe in 2018, the centre left "needs to get a handle on migration" (Wintour 2018). Parties on the left will not only continue to lose

[1] Quoted in Centre for Contemporary Cultural Studies, *EMPIRE STRIKES BACK: Race and Racism In 70's Britain* (Routledge, 2004), 283.

I. Danewid (✉)
Department of International Relations, University of Sussex, Brighton, UK
e-mail: I.Danewid@sussex.ac.uk

G. Proglio et al. (eds.), *The Black Mediterranean*, Mediterranean Perspectives, https://doi.org/10.1007/978-3-030-51391-7_8

votes as long as they stand behind open borders and immigration, Clinton argued; worse, pro-refugee policies are in fact anti-democratic, an imposition on sovereignty, and a threat to social welfare.

Clinton's endorsement of welfare chauvinism is by no means unique. Social democratic parties all over Europe have increasingly come to reject open borders in favour of the very welfare chauvinism they until recently were at pains to brand as far-right and fascist. In some countries, immigration has even ceased to be a left or right issue: in Denmark, border security is effectively a pre-political question, as *all* parties—right and left—support radically restrictive immigration policies (Mirzoeff 2019). As Angela Nagle explains in her essay "The Left Case Against Open Borders", social democrats have for too long blinded themselves from seeing the true cost of immigration. Open-border "radicalism" only "benefits the elites within the most powerful countries in the world, further disempowers organized labour, robs the developing world of desperately needed professionals, and turns workers against workers" (Nagle 2018). Similar arguments have recently been put forward by scholars such as Wolfgang Streeck: progressive politics, he argues, cannot be reconciled with immigration and "open borders with no ceiling" (Streeck 2016).[2]

In this chapter I take these leftist calls for nativist social democracy as a starting point for examining the relationship between race, borders, and global capitalism. I argue that populists and globalists, in spite of their many differences, share an understanding of borders and globalisation as a "zero-sum game" in which the success of one depends on the demise of the other (Friedman 2007). What neither of these perspectives can explain is why globalisation, rather than ushering in the emergence of "a world without borders", actually has gone hand in hand with the hardening and militarisation of borders, especially along the North-South equator. As Wendy Brown (2010: 7–8) explains, "what we have come to call a globalized world harbours fundamental tensions between opening and barricading… These tensions materialise as increasingly liberated borders, on the one hand, and the devotion of unprecedented funds, energies and technologies to border fortification, on the other." From the West Bank to the US-Mexico border, to Kashmir and the Mediterranean, vast amounts of state money has been poured into the construction of walls and other barriers, the policing of borders, and the expansion of the detention

[2] John Judis (2018) similarly argues that "without control of borders and immigration, it is very hard to imagine the United States becoming a more egalitarian society." See also Miller (2016).

estate—all which began well before the rise of far-right, fascist, and populist political parties across the global North (Ibid.; Vallet and David 2012).

To make sense of this process, this chapter draws on the Black radical tradition. It argues that (racialised) mobility controls and capital accumulation are historically and logically connected. European capitalist modernity was since its inception linked to the creation of highly expendable, super-exploitable, and *moveable* labouring subjects, including African captives, sharecroppers, peons, and indentured servitude labourers, among others. In the aftermath of decolonisation and the "racial break", these older forms of direct, colonial control over the global labour supply have gradually been replaced by a system of resource extraction and continued "sweated" labour in the global South, and the creation of "immigrant labour" in the North—a bifurcated system kept in place by the racialised global border regime. As Harsha Walia (2014) explains, "capitalism requires precarious and exploitable workers to facilitate capital accumulation, and creates those precarious lives through hierarchies of systemic oppression along with its extractions of labour and land." Borders are central to this process: by creating differential zones of labour, they naturalise a system built on the hyperextraction of surplus value from racialised bodies. It is ultimately within this context that contemporary calls for closed borders and nativist social democracy must be examined and understood: not as exceptions from an otherwise peaceful European normality, but as part of the long history of racial capitalism which has always sought to control the movement of the poor by "cut[ting] the social fabric at its weakest, i.e. through the bodies of those racialized, gendered and nationalized as undeserving" (Shilliam 2018).

To develop these arguments, this chapter brings the Black Mediterranean into conversation with the literature on racial capitalism. I argue that this produces a new way of understanding the relationship between race, borders, and capitalism, which directly challenges and critiques the starting premise of welfare chauvinists: namely, that open borders and progressive politics are irreconcilable. Reminding us that the category of the European working class exceeds the citizenry and geographical boundaries of Europe, such an analysis also opens up space for new forms of emancipatory politics—centred on abolition rather than hospitality—which connect the struggles against the global sweatshop "abroad" and the exploitation of migrant labour and the (multi-racial) working classes at "home." If, as Bridget Anderson (2017: 1536) has argued, "[t]he contemporary challenge is to draw out the connections between increasing European poverty and associated popular anger and resentment on the one hand, and the deaths and desperation battering at Europe's borders on the other,"[3] then such a perspective has much to offer.

The chapter proceeds in three parts. I begin by reading the history of the Black Mediterranean through the lens of racial capitalism, arguing that the birth and development of the world capitalist system depended on the exploitation of racialised, coerced, and *migratory* labour. As Lydia Potts (1990: 200) has shown, imperial transfers of labour power were central to the development of capitalism: indeed, "the world market for labour power was indivisibly linked with colonialism well into the twentieth century." Building on this, in the second section I argue that mobility controls remain central to global capitalism, in new but old ways. While previous forms of direct, colonial controls of the world market for labour have been replaced by borders and restrictive immigration policies, the production of racialised, migratory, and sweated pools of labour is still essential to the world economy. In the third and final section I explore how this alternative perspective opens up new possibilities for solidarity and emancipatory politics. Linking this to Paul Gilroy's call for a politics of transfiguration, I theorise the Black Mediterranean as a revolutionary counter-narrative which enables the creation of new forms of resistance, internationalism, and relationality: a solidarity anchored in abolition rather than hospitality, and which frames racial capitalism and state violence—rather than migration—as the underlying problem.

"At the Bottom of the British Cuppa": Migration and the Making of Capitalist Modernity

The racial specificity of Europe's migrant "crisis" is rarely acknowledged (De Genova 2017). Indeed, while the risk of border-crossing death is disproportionately inflicted upon migrants and refugees from sub-Saharan Africa, few scholars have theorised this from the perspective of the global colour line. Inspired by Paul Gilroy's *Black Atlantic*, the emerging scholarship on the Black Mediterranean invites us to do precisely this: to think of the migrant "crisis" as a *racial* crisis that must be understood within the context of Europe's constitutive history of empire, colonial conquest, and transatlantic slavery. As Khalil Saucier (2016) explains, what we are

[3] As Anderson (2017: 1536) goes on to ask: "How to build connections between the low waged, homeless and unemployed EU citizens, those struggling to get by, and the struggles of migrants without turning them into competitors for the privileges of membership?"

witnessing today is "a new declination of an older repressed issue" that "has its roots in Mediterranean racial slavery, Enlightenment thought (i.e. humanism that has relied on the provision of a dehumanized other), the colonial North-South relationship, its colonial legacy, as well as in its fascist and imperial worldview." So far there has been no attempt to examine the Black Mediterranean through a distinct focus on the political economy of migration. In what follows, I take up this task: drawing on the literature on racial capitalism, I show that migration is not a new phenomenon brought about by globalisation. Alongside white settler migration to the Americas, Australia, Canada, New Zealand, and South Africa, racialised, coerced, and less-than-free labour migration was central to the construction of European capitalist modernity. It is ultimately within this wider context of racial capitalism that contemporary debates about welfare chauvinism must be situated and understood.

From W.E.B. Du Bois to Cedric Robinson to Ruth Wilson Gilmore, scholars working within the Black radical tradition have long examined the distinctively racial logics of capitalism (Du Bois 1999; Davis 2011; Gilmore 2007; Johnson and Lubin 2017; Melamed 2015; Robinson 2000; Williams 1994). In contrast to orthodox Marxist theory, these thinkers highlight how the history of capitalism began with the slave trade and not with the factory system; in fact, there was never such a thing as capitalism without slavery, and "the history of Manchester never happened without the history of Mississippi" (Johnson 2016). As Lisa Lowe (2015) explains, capital has never expanded by "rendering all labor, resources, and markets across the world identical" but has always operated by "seizing upon colonial divisions, identifying particular regions for production and others for neglect, certain populations for exploitation and still others for disposal."[4] Put differently, capital has never sought to make labour homogenous or interchangeable but has maximised its profits "precisely through the social

[4] Marx, of course, was not unaware of the problems posed by this racial world order. He condemned both colonialism and slavery, and called on workers to oppose racism. Nonetheless, by bracketing racial violence as a form of "so-called primitive accumulation"—and, thus, as something that belongs to a separate historical era—he neglected to interrogate the link between racial difference and the logic of capital. Rather than a process that is integral to capital accumulation, racism, for Marx, was an embarrassing residue of pre-capitalist social relations. Consequently, while Marx condemned colonialism, he ultimately thought that capitalism bore little responsibility for the trade in human bodies, the theft of Indigenous lands and resources, and the colonial genocides committed in the name of Western civilisation.

productions of 'difference'… marked by race, nation, geographical origins, and gender" (Lowe 1996: 27–8). Race, then, is not a separate form of oppression that sometimes intersects with capital accumulation, as some Marxist might suggest. Rather, race-making practices are fundamental to the operation of capital: in the words of Gilmore (2020), "capitalism requires inequality, and racism enshrines it." Hence the term *racial* capitalism.

In recent years, scholars such as Jodi Byrd (2011), Shona Jackson (2012), Barbara Krauthamer (2013), and Lisa Lowe (2015) have built upon these insights to interrogate the global interconnections between various forms of racialised exploitation under European colonialism. Their research demonstrates how, from 1492 onwards, racialised, coerced, and less-than-free migratory labour was central to the rise and development of capitalist modernity. From the forced labour of Native Americans, to the enslavement and coerced migration of Africans, to the export of the Chinese and Indians as coolies, transports of convicts, "blackbirding" in the Pacific, and numerous forms of intra-continental coerced and migrant labour, "many middle passages" (Christopher et al. 2007) built the modern world.[5] Exact numbers for these transfers are difficult to provide, but estimates provide a glimpse of their enormous magnitude. For example, 12.5 million Africans were enslaved and transported across the Atlantic to satisfy the colossal demand for labour on the sugar plantations of the West Indies and in Brazil. Another 11 million are believed to have died during the march to the African West coast or in confinement awaiting the slave ships which would carry them across the Atlantic (The Transatlantic Slave Database). Numbers for the Indian Oceanic slave trade are even less well-known, but estimates suggest that more than 1 million enslaved Africans were taken Mozambique alone (Alpers 2007). After the phasing out of slavery, indentured labour became a common form of labour recruitment: according to Potts (1990: 72–3), from 1835 to 1941 anywhere between 12 and 37 million Indian and Chinese indentured labourers left for Southeast Asia, Hawaii, the West Indies, California, and Australia to work in the European colonies. Between 1860 and 1900, more than 100,000 Pacific Islanders were transported to Australia, Fiji, New Caledonia, and other Pacific locations; from 1787 to 1868, 160,000 British and Irish

[5] As Christopher et al. (2007: 2) explain, "the rise of capitalism from the late sixteenth century onward forced millions of expropriated people to make middle passages from Europe and Africa to the Americas."

convicts were shipped to Australia. Alongside these forms of coerced and less-than-free migrations, the conquest of the New World propelled a large wave of settler colonial migration. From 1850 to 1920, 70 million people left Europe for the Americas, Australia, Canada, New Zealand, and South Africa, settling on land cleared by the genocide of Indigenous populations (Massey 1990).[6] As Encarnación Gutiérrez Rodríguez (2018: 21) concludes, settler colonial "transatlantic European migration [was] foundational to the creation of these nation-states as countries of immigration." In the United States alone, the population grew from under just 4 million in 1798 to 76 million in 1900, largely due to European migration.

Whatever vantage point one chooses—settler colonialism, indentured servitude, the transatlantic slave trade, or another—it is ultimately difficult to avoid the conclusion that migration was anything but central to the creation of the modern world. Indeed, as Cedric Robinson (2000: 23; see also Cohen 2012) explains, "there has never been a moment in modern European history (if before) that migratory and/or immigrant labor was not a significant aspect of European economies." The growth of cities like London, Glasgow, and Liverpool; the expansion of shipping, insurance, and banking; technological inventions such as James Watt's steam engine: these (capitalist) developments all happened after, and as a result of, the profits generated from these peripheries of unfree, racialised labour (Drayton 2011: 105; Williams 1994).

If migratory labour was central to European colonialism at large, the importation of a super-exploitable, racialised workforce was also necessary to meet the labour demands of industrialisation *within* Europe. In 1851 there were already 700,000 disenfranchised Irish migrants living in England; in France there were 379,000 foreigners, a figure which quickly grew to 1,127,000 in 1886 (Cohen 2012). These early foreign recruits predominantly came from Belgium, Spain, Italy, Russia, and Eastern Europe, and typically found work in informal self-employment, manufacturing, and "sweated trades" such as the clothing industry. After the First World War, they were increasingly superseded by labour recruitments from the colonies. Up until 1975, when immigration restrictions were introduced in most Western European countries, large-scale "guest worker" programmes and the arrival of millions of postcolonial migrants

[6] Smith (2016: 108) estimates that the total migratory flow equated to more than one sixth of the 408 million people living in Europe in 1900.

formed an indispensable part of the European labour force.[7] Those who today blame migrants and minorities for undermining the position of Europe's "native" White working class actively forget this history. As Richard Drayton (2012: 162) explains, when in the 1950s "British democracy created a welfare state at home, it too depended on invisible donations from tea-pickers in Ceylon, rubber-tappers in Malaya, goldminers in South Africa, copper-miners in Rhodesia, the oil of Iraq and especially Iran." European social welfare and migration have historically never been antithetical forces. Quite the opposite: European colonialism and the labour provided by racialised "peripheral" and migrant non-citizen labourers were integral to the development of the European welfare state (Bhambra and Holmwood 2018: 5).

To take seriously this history of coerced, racialised, settler, and "peripheral" labour is, therefore, to trouble easy distinctions between European "citizen" and "non-citizen" workers. The category of the European working class has always extended beyond the geographical borders and citizenship of Europe. In fact, and in contrast to E. P. Thompson's (1963) famous argument, the history of the English working class begins not in the late eighteenth century and as a result of industrialisation, but on the sugar plantations of the New World in the 1660s. The issue here is not only that "migration/immigration [was] central to the creation of the world capitalist system" (Robinson 2014: 2) but, more fundamentally, that what we today call "Europe" itself is a product of the coerced and less-than-free labour performed by slaves, coolies, migrants, and other workers transported by empire across the world oceans and continents. In the next section, I argue that these are not only historical questions: racialised, migratory, and sweated labour is in fact central to the contemporary world economy, in new but old ways. As Robin Cohen (2012: 1) has argued, "whether manufacturing is exported to low-wage areas or migrants are imported to work in metropolitan service sectors, the distinctions between established workers, privileged foreigners and helot labourers have remained and may even have deepened."[8] As we shall see, global

[7] According to Cohen (2012: 46), by the mid-1970s when immigration restrictions were introduced in all Western European countries, 13.5 million "foreigners" were officially counted as residing in Belgian, Denmark, France, Germany, Ireland, Netherlands, the UK, Sweden and Switzerland. See also Van Mol and de Valk (2016: 31–55).

[8] As Cohen goes on to explain, "there is much continuity in the evolution of global migration flows, particularly when we observe that large numbers of subordinated workers continue to meet the demand for low-cost production and service provision."

migration not only "continues to follow geographic routes and rely on socioeconomic systems of exploitation that have been in place since the fifteenth century" (Papastergiadis 2018: 60; see also Mayblin 2017). More disturbingly, restrictive migration policies today function as one of the main vectors through which populations racialised as non-White are rendered surplus under racial capitalism.

Borders and the "Racial Break"

The era following World War II is sometimes described as a "racial break" that marked a shift from colonial, segregationist, and apartheid practices and policies towards racial equality and democracy. As the civil rights and national liberation movements gained momentum, the old-world racial order was plunged into crisis. In her important work on the postwar racial break, Jodi Melamed (2011: 84) recounts how

> Previous articulations of race, geopolitics, and capitalism relied on white supremacy to ideologically unify colonialism and its corresponding capitalist relations. After World War II, however, white supremacy entered a phase of permanent crisis, spurred initially by the casting of the war as a fight against racism and fascism and by the numerous and overlapping postwar anticolonial and antiracist movements.

The result was a shift in the historical articulation of race and capital, "with white supremacy and colonial capitalism giving way to racial liberalism and transnational capitalism and, eventually, to neoliberal multiculturalism and globalization" (Melamed 2006: 2). After 1945, overt forms of racism were effectively discredited. Racial privilege nonetheless lived on; it "evolved to take on new forms adapted to postcoloniality and the demise of legal segregation" (Ibid.). Indeed, race remained tied to capital accumulation in complex ways, as the postwar era would continue "to fuse technologies of racial domination with liberal freedoms to represent people who are exploited for or cut off from capitalist wealth as outsiders to liberal subjectivity for whom life can be disallowed to the point of death" (Ibid.).

Borders emerged as one key site for these processes. As older forms of direct, colonial control over the global labour supply were discredited, they were gradually replaced by a system of resource extraction and continued sweated labour in the global South, on the one hand, and the creation of "immigrant labour" in the North, on the other—a bifurcated system anchored in the racialised global border regime. The racial coding of borders, which were to become so characteristic of the post-1945 world, was of course not new. As Rodríguez (2018) shows in her work on the "coloniality of migration", borders were from their inception tied to racialised notions of desirable and undesirable subjects.[9] The first immigration policies emerged "in the nineteenth century as a modern nation-state colonial tool of governing the population in racial, ethnic, national, religious, and cultural terms" (Ibid.: 21). These policies were engineered and implemented in countries emerging from colonial to sovereign rule—in the Americas in the late nineteenth century and Oceania and South Africa in the early twentieth century—where they served to prioritise the recruitment of White European migrants "while banning certain social, national, religious and racial groups from entry" (Ibid.: 22).

The fortification of borders in the post-1945 era continued and generalised this pattern of racially coded migration policies. In the decades following World War II, Europe initially encouraged labour migration and guest worker programmes in an effort to rebuilt its economy. This came to an end in the 1970s, when most Western European states began to introduce restrictive immigration policies. In the UK, Enoch Powell called for an end to migration with his infamous "Rivers of Blood" speech, warning that "in fifteen or twenty years' time the Black man will have the whip hand over the white man." In France, the National Front rose to prominence with the message that "2 million unemployed means 2 million immigrants too many" (Gabriel 2002: 159) While the rise in xenophobia, racism, and restrictive migration policies often are discussed as a result of the economic downturn and high levels of unemployment that followed the 1973–4 oil crisis, they were—as Stuart Hall and his co-authors (2013) argued so brilliantly in *Policing the Crisis*—also a latent response to an underlying crisis in European state hegemony fuelled by decolonisation, the structural transformation of the world economy, and experiments with Keynesian welfare state politics. Starting in the 1960s and 70s, Western companies began to outsource manufacturing production to postcolonial, low-wage countries.

[9] Important work (Luibhéid 2002; van Walsum 2008) suggests that borders also are gendered. In fact, the very language of protecting borders is heavily gendered and sexualised. As Katrina Schlunke (2002) explains, borders are typically described as "vulnerable shores that must be kept intact and secured against the threat of un-negotiated penetration by strangers."

While restrictive migration policies often are discussed in isolation from this trend of global outsourcing, they are two sides of the same coin. Indeed, mobility controls are central to the globalisation of capital because they enable the hyperextraction of surplus value from racialised populations. By preventing the South's surplus populations from emigrating, borders work to artificially create vast wage differentials between the South and North. Borders thus enable a system where multinational corporations are heavily incentivised to outsource production to the cheapest places possible—which is why 83 percent of the world's manufacturing workforce today lives and works in the global South (Smith 2016: 102; see also Jones 2016; Mezzadri 2016, Muñoz 2016; Seabrook 2014). As John Smith (2016: 104–5) explains, "[j]ust as the infamous pass-laws epitomized apartheid in South Africa, so do immigration controls form the lynchpin of an apartheid-like global economic system that systematically denies citizenship and basic human rights to the workers of the South and which, as in apartheid-era South Africa, is a necessary condition for their super-exploitation." In other words, borders naturalise a system built on the exploitation of hundreds of millions of workers throughout the global South. Caught in what Mike Davis refers to as the "planet of slums", they are "the source of surplus value sustaining profits and feeding unsustainable overconsumption" (Ibid.: 10) in the global North.

In spite of the last few decades' rapid growth in restrictive immigration policies, more people than ever before are on the move. There are now more than 258 million international migrants in the world, a figure that has doubled since the 1980s (International Migration Report 2017). This does not necessarily mean that borders are failing. Indeed, while migration scholars sometimes have been prone to conceptualise borders as physical walls that clearly separate the inside from the outside, borders are far less rigid than this literature makes them out to be. As Bridget Anderson (2017: 1532) explains, borders are at once devices of exclusion *and* inclusion: they are not simply "taps that attempt (successfully or not) to control the flows of entry of non-citizens" but also function as "moulds that shape social relations."[10] Borders lock undocumented migrants into a state of permanent precarity, vulnerability, and super-exploitability—and thus cre-

[10] As Mezzadra and Neilson (2013: 7) explain, borders are "equally devices of inclusion that select and filter people and different forms of circulation in ways no less violent than those deployed in exclusionary measures" – "inclusion existing in a continuum with exclusion, rather than in opposition to it."

ate a non-citizen workforce which lacks access to civil and social rights (such as unemployment benefits, social security, housing allowance, tax credits, pensions): a segment of the labour force which Peter Nyers (2003: 1070; see also Cross 2013; de Genova 2013; Mitropoulos and Kiem 2015) refers to as the "deportspora." As Harsha Walia (2014) makes clear, this state-mediated "in-gathering of off-shore labor" is a pillar of contemporary neoliberalism, because it allows for the accumulation of domestic capital. Borders ultimately produce a division between "citizen" and "non-citizen" workers, the latter constituting a commodified, exploitable, flexible, and expendable workforce, "a subordinate reserve army of deportable 'foreign' labour, always-already within the space of the nation-state, readily available for deployment as the inevitably over-employed working poor" (de Genova 2013: 1190).

Consider the following examples: In Italy, exploitation of migrant labour is widespread in the agricultural sector. Kojo, a migrant worker interviewed by Amnesty International, recounts how "We were working from 6 in the morning to 6 in the evening, every day of the week, for 20 euros a day. We could not take breaks, not even for eating. We used to eat the oranges on the trees" (Amnesty International 2012: 7). In the UK, migrants held in detention centres are excluded from the national minimum wage. Performing the menial but vital tasks necessary to keep the centres operating smoothly (cleaning, cooking, and so on) detainees work for as little as £1 an hour, saving the private corporations running the centres up to £1.5 million per year (Bales and Mayblin 2018). In Germany, asylum seekers are regularly hired to "one-euro jobs" for which they earn a monthly income of roughly €145 (The Local 2016). Examples are plentiful. Indeed, these cases are not exceptional but representative of a larger system which relies on borders and the criminalisation of migration to produce a super-exploited, hyper-surveilled, and expendable labour pool. This is not a marginal phenomenon; according to the ILO (2018), there are currently 164 million migrant workers in the world. Borders play a central role in enforcing this global system of apartheid. As Fernandez et al. (2006: 467) explain, "Without the border, there would be no differential zones of labor, no spaces to realize surplus capital through the dumping of overproduction, no way of patrolling surly populations that might want to resist proletarianization, no release valve for speculative access."

In sum, half a decade after the postwar "racial break", racialised, migratory, and sweated labour remains central to the world economy—in new but old ways. Today's global border regime not only serves to produce new pools of low-wage labour, but is arguably a direct descendant of older forms of direct, colonial control over the global labour supply. In the words of Melamed (2006: 1), "Race continues to permeate capitalism's economic and social processes, organizing the hyperextraction of surplus value from racialized bodies and naturalizing a system of capital accumulation that grossly favours the global North over the global South." Borders are central to this system because they at once enable the exploitation of sweated labour (in the global South) and migrant labour (in the North). While production outsourcing and migration often are studied separately, they are thus more fruitfully understood as interlinked aspects of the same "wage-differential-driven transformation of global production" (Smith 2016: 44). As one of the main routes through which populations racialised as non-White are rendered surplus under contemporary racial capitalism, the global border regime has effectively replaced earlier forms of direct, colonial controls of the global labour supply.

To take this insight seriously is not only to trouble the dichotomy between the European "White working class" and migrants, as it demonstrates that any discussion of the European working class—*then* as well as *now*—has to extend beyond border regimes, citizenship, and colour lines. More than that, and in sharp contrast to contemporary calls for nativist social democracy, it also makes clear that there can be no progressive, anti-capitalist case for borders. As Anderson (2017: 1627) explains, "[a] new politics of migration must make connections between migrants and citizens" as well as "between migration and other global processes, particularly outsourcing and the exploitation of labour and resources in the global South." In the next section, I look at how this alternative understanding of the relationship between migratory labour, mobility controls, and European capitalist modernity opens up new possibilities for anti-capitalist, anti-racist solidarity—a solidarity grounded, not in hospitality towards migrants, but in the abolition of borders.

"These Walls Must Fall": From Hospitality to Abolition

Writing about the Black Atlantic, Paul Gilroy was always quick to remind his readers to think of it as more than a space of enslavement, exploitation, and racial terror. Indeed, the Black Atlantic was also a counterculture of resistance with the power "to conjure up and enact the new modes of friendship, happiness, and solidarity" (Gilroy 1993: 38). Where a politics of fulfilment is "content to play occidental rationality at its own game" (Ibid.: 37) the Black Atlantic opened up space for a politics of transfiguration: that is, "the emergence of qualitatively new desires, social relations, and modes of association within the racial community of interpretation and resistance *and* between that group and its erst-white oppressors" (Ibid.). The Black Mediterranean, I would like to suggest, similarly offers resources for envisioning new modes of resistance, rebellion, and relationality. As Robbie Shilliam (2015: 183) writes,

> If we think differently about the past then different possibilities open up for our present. Instead of an imperial world map, we might begin to glean a global infrastructure of anti-colonial connectivity; and rather than reproducing categories of segregation we might start to live an ethos of humanity, deeply relating.

When read through the lens of the Black radical tradition, the Black Mediterranean demonstrates the importance of rearticulating the struggle for migrant rights as a struggle against racial capitalism and state violence: a struggle for abolition rather than hospitality. This can be contrasted with liberal, state-centric framings of migrant justice which typically fail to question the statist logics that make "immigration" an issue in the first place, and which therefore maintain structures of subjugation beneath the rhetoric of rights, reconciliation, and formal equality. As Naomi Paik (2017: 16) explains, liberal perspectives anchored in hospitality problematically shore up "the notion that undocumented immigrants deserve inclusion in the community, but contingent on their submission to the capitalist extraction of their labour and to the state's (racialised) criminal justice apparatuses." In place of hospitality, the Black radical tradition shifts our focus to what W.E.B. Du Bois (1999: 16; see also Davis 2005) referred to as "abolition democracy": a politics which seeks to abolish all institutions complicit in racial capitalism. The border, I have argued, is one of these.

A number of activist groups and movements within Europe are already practising such an abolitionist politics, and from them there is much to be learned.[11] Calling for a transfigurative anti-capitalist, anti-racist politics of No Borders, they not only seek to break away from border-bound definitions of citizenship by challenging the very distinction between "citizen" and "non-citizen." More than that, they also unravel how the European border regime is intimately entwined with a set of systems of power and repression—at once at "home" and "abroad"—which find their roots in colonisation, slavery, and capital accumulation and exploitation (Walia 2014). By linking the mass drownings of migrants in the Mediterranean to ongoing struggles for justice *within* Europe, they thus invite us to view the struggles of migrants, minorities, workers, detention detainees, and the undocumented within a shared circuit: different fronts of the same war. As Walia (2014: 39) explains, such solidarity "encapsulates a dual critique of Western state building within global empire: the role of Western imperialism in dispossessing communities in order to secure land and resources for state and capitalist interests, as well as the deliberately limited inclusion of migrant bodies into Western states through processes of criminalization and racialization that justify the commodification of their labor." To flesh out these arguments, in what follows I focus on two movements which I argue are anchored in abolition rather than hospitality: the protests against Yarl's Wood in the UK, and *Parti des Indigènes de la République* (PIR) in France.

In the UK, the #HungerForFreedom strikes at Yarl's Wood immigration removal centre have served as a rallying point for activists and organisers calling for an end to the deportation of migrants and the UK government's hostile environment policy. The UK has one of the largest detention systems in Europe, confining up 30,000 people (including children) every year. It is also the only country in Europe with no time limit on detention. Since it opened in 2001, Yarl's Wood has been subject to repeated criticism of racism, sexual abuse, and inadequate mental and physical healthcare (Travis 2017). The government's own Shaw Report (2016) called for a complete ban on the detention of pregnant women, and strongly advised against detention of victims of rape and sexual violence as well as those with post-traumatic stress disorder—all of which are common at Yarl's Wood. The centre has, in spite of this, remained opened. In February 2018, more than 100 women detained at Yarl's Wood went

[11] Amongst numerous others, see the No Border Network, Black Lives Matter UK, Parti des Indigènes de la République, New Urban Collective, European Network Against Racism, and Campaign Against Police and State Violence.

on a hunger strike. More than 2000 protestors—spearheaded by a network of migration justice activists and grassroots groups, including Detention Action, Freed Voices, Movement for Justice, SDS, and Right to Remain—gathered around the centre to echo their demands (Townsend 2016). Self-consciously abolitionist and transfigurative, organisers are not simply calling for better conditions for those detained at Yarl's Wood and elsewhere. Rather, they are demanding an end to detention altogether, as well as amnesty for all those who arrived in the country as children and those who have been in the UK for more than 10 years. In light of this it is perhaps no coincidence that the movement's rallying cry is a direct reference to the Fallist movements in South Africa: "These Walls Must Fall."

In France, the *Parti des Indigènes de la République* (PIR) has been at the forefront in drawing attention to the continuity between mass migrant deaths during Mediterranean crossings and the everyday violence inflicted on racialised minorities *within* Europe. PIR was formed in early 2005 with the goal of contributing to "the emergence of a political and organized expression of the rage of immigrant populations."[12] PIR is primarily composed of French youths of African, Arab, Muslim, Maghrebian and Antillean origin, born and raised in France, who live the experience of racism, marginalisation, and exploitation. The French term *indigéne*—loosely translated as Indigenous—was chosen by the group to invoke the colonial populations who, up until 1946, were governed by the *Code de l'Indigénat*. The notion of *indigéne* thus draws attention to the fact that the French Republic—while claiming to uphold colour-blind values of equality, fraternity, and liberty—in fact continues to treat some of its citizens as quasi-colonial subjects. As Horuia Bouteldja, the spokesperson of PIR, explains:

> When they refuse to accept us as French citizens, they deny us equality. We need to name this reality: we cannot be French, so we are native. We are second-class citizens; ours is a lumpen-citizenship, just as at the time of the colonies. This imaginary linked to colonization and the history of slavery continues to determine how they perceive us, for the body of the Indigenous was constructed during the colonial era. As long as this imaginary is alive, we remain native.[13]

[12] Sadri Khiari, quoted in Kipfer (2011: 1158).
[13] Quoted in Ibid.

For PIR, the legacies of race and colonialism continue to structure present realities in Europe: from the dark waters of the Mediterranean, where migrants are left to die, to Yarl's Wood, camps for agricultural migrant workers in Italy, and the *banlieues* of Paris, the violence of racial capitalism cuts straight through Europe.

In linking together metropolitan cores and colonial peripheries, movements such as PIR and the #HungerForFreedom strikes open up space for a different form of solidarity, anchored in abolition rather than hospitality. They do this by targeting three specific problems: first, the problem of racism, stigmatisation, and marginalisation within Europe; second, the role of racism in creating a super-exploitative work force both within and outside of Europe; and third, the necessity of linking anti-racist struggle within Europe to internationalist liberation struggles worldwide (Kipfer 2011: 1162). As Prem Rajaram (2018: 11) explains, this leads to a perspective which "allows for thinking points of articulation between different marginalised groups, refuses the state-centric account of migrant and refugee 'governance,' and allows also for the basis of a politics of solidarity." This is a solidarity which goes beyond liberal integrationist visions based on Western generosity and pathways to citizenship. Rather than "a liberal way of dealing with migration as a temporary crisis that can be managed with the likes of the EU–Turkey migrant deal" (Turhan and Armiero 2017: 7), this is a transfigurative and abolitionist politics that rearticulates the struggle for migrant justice as a struggle against racial capitalism and state violence.

Conclusion

In this chapter, I have argued that the Black Mediterranean offers resources for rewriting the history of European capitalist modernity in ways that challenge the popular distinction between "immigrants" and the "White working class." Drawing on the Black radical tradition, I have argued that there can be no anti-capitalist, anti-racist case for closed borders, because the history of capitalism *is* a history of the production, dispossession, and exploitation of racialised, migratory labour. The creation of a highly expendable, super-exploitable, and moveable workforce (including captives, sharecroppers, and indentured servants) was historically central to the world capitalist system. Today, under neoliberalism and corporate globalisation, previous forms of direct, colonial control of the global labour supply have been replaced by the creation of a system of resource extraction and sweated labour in the global South and the production of migrants as vulnerable, deportable, and therefore super-exploitable workers in the North—a system kept in place by the racialised global border regime. Mobility and its control have in other words always been central to capitalism, in history as well as in the present. As PIR and the #HungerForFreedom strikes remind us, this means that the struggle for migrant justice is a struggle for abolition—and not for hospitality. There can be no emancipatory struggle worthy of its name which is not simultaneously a struggle *against* borders. As organisers within and beyond Europe remind us, "These Walls Must Fall."

References

Alpers, Edward. 2007. The Other Middle Passage: The African Slave Trade in the Indian Ocean. In *Many Middle Passages: Forced Migration and the Making of the Modern World*, ed. Emma Christopher, Cassandra Pybus, and Marcus Rediker, 1st ed. Berkeley: University of California Press.

Amnesty International. 2012. Exploited Labour: Migrant Workers in Italy's Agricultural Sector. https://www.amnesty.org/download/Documents/20000/eur300202012en.pdf

Anderson, Bridget. 2017. Towards a New Politics of Migration? *Ethnic and Racial Studies* 40 (9): 1527–1537. https://doi.org/10.1080/01419870.2017.1300297.

Bales, Katie, and Lucy Mayblin. 2018. Unfree Labour in Immigration Detention: Exploitation and Coercion of a Captive Immigrant Workforce. *Economy and Society* 47 (2): 191.

Bhambra, Gurminder, and John Holmwood. 2018. Colonialism, Postcolonialism and the Liberal Welfare State. *New Political Economy* 23: 5.

Broeck, Sabine, and P. Khalil Saucier. 2016. A Dialogue: On European Borders, Black Movement, and the History of Social Death. *Black Studies Papers* 2 (1): 25–28.

Brown, Wendy. 2010. *Walled States, Waning Sovereignty.* Cambridge, MA: MIT Press.

Byrd, Jodi A. 2011. *The Transit of Empire: Indigenous Critiques of Colonialism.* Minneapolis: University of Minnesota Press.

Centre for Contemporary Cultural Studies. 2004. *EMPIRE STRIKES BACK: Race and Racism in 70's Britain.* London: Routledge.

Christopher, Emma, Cassandra Pybus, and Marcus Rediker, eds. 2007. *Many Middle Passages: Forced Migration and the Making of the Modern World.* 1st ed. Berkeley: University of California Press.

Cohen, Robin. 2012. *Migration and Its Enemies: Global Capital, Migrant Labour and the Nation-State*. Ashgate Publishing.

Cross, Hannah. 2013. *Migrants, Borders and Global Capitalism: West African Labour Mobility and EU Borders*. London: Routledge.

Davis, Angela Y. 2005. *Abolition Democracy: Beyond Empire, Prisons, and Torture*. 1st ed. New York: Seven Stories Press.

———. 2011. *Women, Race, & Class*. New York: Knopf Doubleday Publishing Group.

De Genova, Nicholas. 2013. Spectacles of Migrant 'Illegality': The Scene of Exclusion, the Obscene of Inclusion. *Ethnic and Racial Studies* 36 (7): 1180–1198. https://doi.org/10.1080/01419870.2013.783710.

———. 2017. The 'Migrant Crisis' as Racial Crisis: Do Black Lives Matter in Europe? *Ethnic and Racial Studies*: 1–18. https://doi.org/10.1080/01419870.2017.1361543.

Drayton, Richard. 2011. The Collaboration of Labour: Slaves, Empires, and Globalizations in the Atlantic World, C. 1600–1850. In *Globalisation in World History*, ed. A.G. Hopkins. Random House.

———. 2012. Imperial History and the Human Future. *History Workshop Journal* 74 (1): 156–172.

Du Bois, W.E.B. 1999. *Black Reconstruction in America 1860–1880*. Simon and Schuster.

Fernandez, Carlos, Meredith Gill, Imre Szeman, and Jessica Whyte. 2006. Erasing the Line, or, the Politics of the Border. *Ephemera* 6 (4): 466–483.

Friedman, Thomas L. 2007. *The World Is Flat: The Globalized World in the Twenty-First Century*. 2nd Rev ed. London: Penguin.

Gabriel, John. 2002. *Whitewash: Racialized Politics and the Media*. Routledge.

Gilmore, Ruth Wilson. 2007. *Golden Gulag: Prisons, Surplus, Crisis, and Opposition in Globalizing California*. Berkeley: University of California Press.

———. 2020. *Geographies of Racial Capitalism with Ruth Wilson Gilmore. An Antipode Foundation film*. https://www.youtube.com/watch?v=2CS627aKrJI

Gilroy, Paul. 1993. *The Black Atlantic: Modernity and Double Consciousness*. London: Verso.

Hall, Stuart, Chas Critcher, Tony Jefferson, John Clarke, and Brian Roberts. 2013. *Policing the Crisis: Mugging, the State and Law and Order*. Basingstoke: Macmillan International Higher Education.

ILO. 2018. ILO Global Estimates on International Migrant Workers, December 5. https://www.ilo.org/global/publications/books/WCMS_652001/lang%2D%2Den/index.htm

International Migration Report. United Nations, 2017. http://www.un.org/en/development/desa/population/migration/publications/migrationreport/docs/MigrationReport2017_Highlights.pdf

Jackson, Shona N. 2012. *Creole Indigeneity: Between Myth and Nation in the Caribbean*. Minneapolis: University of Minnesota Press.
Johnson, Gaye Theresa, and Alex Lubin. 2017. *Futures of Black Radicalism*. London/Brooklyn: Verso Books.
Johnson, Walter. 2018. To Remake the World: Slavery, Racial Capitalism, and Justice. Boston Review. http://bostonreview.net/forum/walter-johnson-toremakethe-world.
Jones, Reece. 2016. *Violent Borders: Refugees and the Right to Move*. London: Verso.
Judis, John B. 2018. *The Nationalist Revival: Trade, Immigration, and the Revolt Against Globalization*. Columbia. Global Reports.
Kelley, Robin D.G. 2000. Introduction. In *Black Marxism: The Making of the Black Radical Tradition*, ed. Cedric J. Robinson. Chapel Hill: University of North Carolina Press.
Kipfer, Stefan. 2011. Decolonization in the Heart of Empire: Some Fanonian Echoes in France Today. *Antipode* 43 (4): 1155–1180. https://doi.org/10.1111/j.1467-8330.2011.00851.x.
Krauthamer, Barbara. 2013. *Black Slaves, Indian Masters: Slavery, Emancipation, and Citizenship in the Native American South*. Chapel Hill: University of North Carolina Press Books.
Lowe, Lisa. 1996. *Immigrant Acts: On Asian American Cultural Politics*. Durham: Duke University Press.
———. 2015. *The Intimacies of Four Continents*. Durham: Duke University Press.
Luibhéid, Eithne. 2002. *Entry Denied: Controlling Sexuality at the Border*. Minneapolis: University of Minnesota Press.
Massey, Douglas S. 1990. The Social and Economic Origins of Immigration. *The ANNALS of the American Academy of Political and Social Science* 510 (1): 60–72. https://doi.org/10.1177/0002716290510001005.
Mayblin, Lucy. 2017. *Asylum After Empire: Colonial Legacies in the Politics of Asylum Seeking*. Lanham: Rowman & Littlefield International.
Melamed, Jodi. 2006. The Spirit of Neoliberalism: From Racial Liberalism to Neoliberal Multiculturalism. *Social Text* 24 (4(89)): 1–24. https://doi.org/10.1215/01642472-2006-009.
———. 2011. Reading Tehran in Lolita: Making Racialized and Gendered Difference Work. In *Strange Affinities: The Gender and Sexual Politics of Comparative Racialization*. Durham: Duke University Press.
———. 2015. Racial Capitalism. *Critical Ethnic Studies* 1 (1): 76–85. https://doi.org/10.5749/jcritethnstud.1.1.0076.
Mezzadra, Sandro, and Brett Neilson. 2013. *Border as Method, or, the Multiplication of Labor*. Durham/London: Duke University Press.
Mezzadri, Alessandra. 2016. Class, Gender and the Sweatshop: On the Nexus between Labour Commodification and Exploitation. *Third World Quarterly* 37 (10): 1877–1900.

Miller, David. 2016. *Strangers in Our Midst*. Cambridge, MA: Harvard University Press.

Mirzoeff, Nicholas. 2019. 'Social Death' in Denmark. *The Nation*, January 20. https://www.thenation.com/article/denmark-refugees-asylum-europe/

Mitropoulos, Angela, and Matthew Kiem. 2015. Cross-Border Operations. *The New Inquiry* (Blog), November 18. https://thenewinquiry.com/cross-border-operations/

Muñoz, Carolina Bank. 2016. *Transnational Tortillas: Race, Gender, and Shop-Floor Politics in Mexico and the United States*. Ithaca: Cornell University Press.

Nagle, Angela. 2018. The Left Case Against Open Borders. *American Affairs Journal*. https://americanaffairsjournal.org/2018/11/the-left-case-against-open-borders/

Nyers, Peter. 2003. Abject Cosmopolitanism: The Politics of Protection in the Anti-Deportation Movement. *Third World Quarterly* 24 (6): 1069–1093.

Paik, Naomi. 2017. Abolitionist Futures and the US Sanctuary Movement. *Race & Class* 59 (2): 3–25.

Papastergiadis, Nikos. 2018. *The Turbulence of Migration: Globalization, Deterritorialization and Hybridity*. New York: John Wiley & Sons.

Potts, Lydia. 1990. *The World Labour Market: A History of Migration*. London: Zed Books.

Rajaram, Prem Kumar. 2018. Refugees as Surplus Population: Race, Migration and Capitalist Value Regimes. *New Political Economy* 23 (5): 627–639. https://doi.org/10.1080/13563467.2017.1417372.

Robinson, Cedric J. 2000. *Black Marxism: The Making of the Black Radical Tradition*. 2nd ed. Chapel Hill: University North Carolina Press.

Robinson, William I. and Santos, Xuan. 2014. Global Capitalism, Immigrant Labor, and the Struggle for Justice. Class, Race and Corporate Power 2 (3). https://doi.org/10.25148/CRCP.2.3.16092122

Rodríguez, Encarnación Gutiérrez. 2018. The Coloniality of Migration and the 'Refugee Crisis': On the Asylum-Migration Nexus, the Transatlantic White European Settler Colonialism-Migration and Racial Capitalism. *Refuge* 34 (1): 16–28.

Roediger, David. 2017. *Class, Race and Marxism*. London: Verso Books.

Schlunke, Katrina. 2002. Sovereign Hospitalities? *Borderlands* 2 (1). http://www.borderlands.net.au/vol1no2_2002/schlunke_hospitalities.html

Seabrook, Jeremy. 2014. *The Song of the Shirt: Cheap Clothes Across Continents and Centuries*. New Delhi: Navayana.

Shaw, Stephen. 2016. Review into the Welfare in Detention of Vulnerable Persons. Home Office. https://www.gov.uk/government/publications/review-into-the-welfare-in-detention-of-vulnerable-persons

Shilliam, Robbie. 2015. *The Black Pacific: Anti-Colonial Struggles and Oceanic Connections*. London: Bloomsbury Publishing.

———. 2018. Race and the Undeserving Poor. *The Disorder of Things* (blog). https://thedisorderofthings.com/2018/06/27/race-and-the-undeserving-poor/

Smith, John. 2016. *Imperialism in the Twenty-First Century: Globalization, Super-Exploitation, and Capitalism's Final Crisis*. New York: New York University Press.

Streeck, Wolfgang. 2016. Why Europe Can't Function as It Stands, November 7. https://www.versobooks.com/blogs/2926-wolfgang-streeck-why-europe-can-t-function-as-it-stands

The Local. 2016. Germany Puts Refugees to Work … for One Euro. *The Local*, May 16. https://www.thelocal.de/20160516/germany-puts-refugees-to-work-for-one-euro

The Transatlantic Slave Database. Available at http://www.slavevoyages.org

Thompson, Edward Palmer. 1963. *The Making of the English Working Class*. London: IICA.

Townsend, Mark. 2016. Yarl's Wood Demonstration Draws up to 2,000 Campaigners. *The Guardian*. https://www.theguardian.com/world/2016/dec/03/yarls-wood-demonstration-brexit-vote

Travis, Alan. 2017. More Rape and Torture Victims Being Held at Yarl's Wood, Report Says. *The Guardian*, November 15. https:/j+/www.theguardian.com/uk-news/2017/nov/15/yarls-wood-chief-inspector-prisons-report-rape-torture

Turhan, Ethemcan, and Marco Armiero. April 3, 2017. Cutting the Fence, Sabotaging the Border: Migration as a Revolutionary Practice. *Capitalism Nature Socialism* 28 (2): 1–9. https://doi.org/10.1080/10455752.2017.1315245.

Vallet, Élisabeth, and Charles-Philippe David, eds. 2012 The (Re)Building of the Wall in International Relations. *Journal of Borderlands Studies* 27: 111–119.

Van Mol, Christof, and Helga de Valk. 2016. Migration and Immigrants in Europe: A Historical and Demographic Perspective. In *Integration Processes and Policies in Europe: Contexts, Levels and Actors*, IMISCOE Research Series, ed. Blanca Garcés-Mascareñas and Rinus Penninx, 31–55. Cham: Springer International Publishing. https://doi.org/10.1007/978-3-319-21674-4_3.

Walia, Harsha. 2014. *Undoing Border Imperialism*. Edinburgh: AK Press.

van Walsum, Sarah. 2008. *The Family and the Nation: Dutch Family Migration Policies in the Context of Changing Family Norms*. Newcastle upon Tyne: Cambridge Scholars Publishing.

Williams, Eric (1994). *Capitalism and Slavery*. Chapel Hill: The University of North Carolina Press.

Wintour, Patrick. 2018. Hillary Clinton: Europe Must Curb Immigration to Stop Rightwing Populists. *The Guardian*, November 22. https://www.theguardian.com/world/2018/nov/22/hillary-clinton-europe-must-curb-immigration-stop-populists-trump-brexit

PART III

Citizenship

CHAPTER 9

L'Italia Meticcia? The Black Mediterranean and the Racial Cartographies of Citizenship

Camilla Hawthorne

> If we think of the Mediterranean *not* as the centre of an 'us' surrounded by difference, imperfection…if we think of it instead as a tableau of differences, exchanges, and inheritances, we will begin to see the global dimensions of thought itself. (Gnisci 1998, p. 48)

> [P]eople of African ancestry with Italian citizenship are also routinely perceived as permanent cultural outsiders…The sense that Africans cannot possibly be Italian and are therefore socially peripheral at best is experienced regularly. (Merrill 2018)

> I am an Italian with black skin, not a black woman who feels Italian. (Affricot 2015)

C. Hawthorne (✉)
Sociology and Critical Race and Ethnic Studies, University of California, Santa Cruz, Santa Cruz, CA, USA
e-mail: camilla@ucsc.edu

G. Proglio et al. (eds.), *The Black Mediterranean*, Mediterranean Perspectives, https://doi.org/10.1007/978-3-030-51391-7_9

Locating the Black Mediterranean in Italian Citizenship Struggles

Over the last decade, the Black Mediterranean has emerged as a powerful framework through which scholars have critically engaged in issues pertaining to refugees and migrants crossing the Mediterranean into southern Europe. Working in this vein, geographers, anthropologists, political theorists, sociologists, historians, and literature scholars have employed the Black Mediterranean to address subjects such as the cultural production of the Black diaspora in southern Europe (Di Maio 2012); the lived experiences of Blackness and Black political struggle in Italy (Hawthorne 2017); the racial politics of migration and the ongoing refugee "crisis" (Proglio 2018; Raeymaekers 2015a, b); and histories of racial exclusion or preclusion in the Mediterranean Basin (Danewid 2017; Saucier and Woods 2014). This scholarship represents a necessary corrective to the dominant trends in critical migration and border studies in Europe, which have tended to ignore the question of race entirely or treat it as a mere superstructural embellishment atop the material contradictions of capital (Danewid 2017; Saucier and Woods 2014). Yet comparatively few scholars have used the Black Mediterranean to think through the experiences of Black youth who grew up in Italy and their struggles to gain Italian citizenship.

There are several possible reasons for the elision of Black citizenship struggles in Italy among the growing (and largely Italy-focused) scholarship on the Black Mediterranean. The Black Mediterranean emphasizes the transnational, border-transgressing linkages connecting southern Europe and Africa. Second-generation Black youth (the children of African immigrants), on the other hand, have not emigrated from a foreign country and are in fact seeking recognition as legitimate citizens by the Italian nation-state. There has been a tendency among some scholars to dismiss the mobilizations of the children of immigrants in Italy as insufficiently radical because they are seemingly oriented towards a liberal politics of state recognition. In their place, the abstracted figure of the Black refugee crossing the Mediterranean is vaunted as a symbol for a radical, outernational politics of refusal or fugitivity.[1]

[1] In one telling incident, Giorgio Agamben (who, it should be noted, does not engage with the Black Mediterranean in his work) published a controversial blog post in 2017 explaining that he did not sign a petition in favour of *jus soli* for the children of immigrants in Italy

I wish to argue, however, that the Black Mediterranean can actually serve as a powerful way to think about the *relationship* between refugee rights mobilizations and struggles over citizenship. Analysing these stories together (and engaging with the processes by which they are subsequently pulled apart in both political discourse and in activist praxis) allows us to push back against the dangerous tendency to reproduce racialized hierarchies that distinguish between Black refugees and Black citizens (or citizens-in-waiting). And it is important to remember that the Black Mediterranean in part builds upon (and responds to the limitations[2] of) the Black *Atlantic*, a concept developed by Robert Farris Thompson (Thompson 2010) and Paul Gilroy (Gilroy 1993) to describe the trans-Atlantic cultural formations and political practices of the Black diaspora. In particular, the Black Atlantic for Gilroy was not necessarily intended to serve as an intervention into migration studies, but rather as a conceptual toolkit for Black Britons (and Black Europeans broadly) to address the double consciousness involved in being simultaneously Black and European. In other words, the Black Mediterranean—like the Black Atlantic—can serve as a powerful framework for thinking regionally and globally about the *entanglements* of race, cross-border mobilities, and citizenship. By shifting the geographic centre of gravity, the Black *Mediterranean* permits a careful engagement with the specific histories and geographies of racial formation in the Mediterranean—in particular, how southern European racial liminality has shaped the conditions of possibility for Black struggles across the region.

Chapter Outline

This chapter explores debates unfolding in Italy over the racial boundaries of Italian citizenship through the framework of the Black Mediterranean, drawing on ethnographic and archival research conducted in Italy between

because he believed that citizenship (or, by his description, involuntary inscription into the regulatory apparatus of a nation-state) is the *problem,* not the solution, in the question of exclusion (Agamben 1998, 2017). Of course, as many commentators subsequently pointed out, Agamben did not renounce his *own* citizenship in solidarity with the nearly one million disenfranchised youth of colour in Italy.

[2] For generative critiques of Gilroy's *Black Atlantic,* see Stefan Helmreich's article "Kinship, Nation, and Paul Gilroy's Concept of Diaspora," *Diaspora* 2.2 (1992): 243–249; and Jacqueline Nassy Brown's monograph *Dropping Anchor, Setting Sail: Geographies of Race in Black Liverpool* (Princeton: Princeton University Press, 2005).

2012 and 2017. The deep entanglements of race and citizenship signalled in ongoing debates about the *seconda generazione*, *jus soli*, and Black Italianness are inflected by a long history of anxiety and contestation over the racial status of Italy, perched as it is at the geographical and metaphorical edges of Europe and Africa. I thus begin the chapter by discussing the significance of the Black Mediterranean for analyses of Italian citizenship struggles, constructing a counter-history of Italian citizenship law that foregrounds contestations over the liminal, Mediterranean racial status of Italians.

Questions of who truly "belongs" in Italy as a rightful citizen continue to be shaped by the much-hyperbolized threat of African contamination and Italian racial degeneration. As one illustrative example, I recount the widely publicized story of the forty-two palm trees that were planted in Milan's Piazza del Duomo in 2017. This event proved to be a powerful incitement to discourse (Foucault 1990) that threw into question the boundaries of Italy in relation to the Mediterranean and Blackness, as well as the dynamic forces of globalization, neoliberal capitalism, and mass migration. The debates incited by this urban greening project dovetailed with political contestations unfolding contemporaneously over the rights of refugees and the expansion of citizenship to the children of immigrants born and raised in Italy. In this context, I argue, the Black Mediterranean stages a critical intervention vis-à-vis the resurgence of interest in Mediterranean interconnections as an idealistic, cosmopolitan response to far-right, ethnic absolutist (Gilroy 1993) discourses and restrictive Italian migration/citizenship policies.

In the final section of this chapter, I discuss the implications of the Black Mediterranean not only as a theoretical framework for understanding citizenship struggles, but also as a radical form of political praxis. Drawing on Devon Carbado's notion of *racial naturalization* (Carbado 2005), I argue that while Black citizens-in-waiting and Black refugees are differentially positioned in relation to the Italian state's "racializing juridical assemblages" (Weheliye 2014, p. 79), they share the condition of being simultaneously inside and outside of a precariously white,[3] Mediterranean

[3] In this chapter, I capitalize "Black" to suggest, if not direct ties of shared ancestry, shared histories of subjugation and struggle that produce historically- and geographically-specific articulations of Blackness as a political collectivity and subjectivity. In contrast, I do not capitalize "white." This is because my aim is to denaturalize "whiteness" as a unitary category by showing the ways in which its reproduction and shifting boundaries function to regulate access to personhood, privilege, property, and rights—including citizenship (Harris 1993).

Italian nation. Yet the emergent, transgressive solidarities enabled by a Black Mediterranean diasporic ethics imagine political possibilities beyond the nation-state and its practices of exclusion. They cultivate relationships of care based on shared histories of struggle rather than on blood, birthplace, or legal status. Finally, I conclude with some reflections on how the Black Mediterranean, by provincializing the Black Atlantic, can illuminate the insidious entanglements of colonialism, neoliberal migration management, and racial citizenship that are increasingly shaping the terrain of possibility for Black activism across the modern world.

Mediterraneanism, *Meticciato*, and the Making of Italians

The ambiguous trans-Mediterranean intimacies and intersections between Italy and Africa, and between Blackness and the precarious whiteness of Italianness, have shaped my own lived experience as a self-identified Black Italian—the daughter of a white Italian[4] mother and a Black American father who grew up between California and northern Italy. I have Italian citizenship thanks to the Italian "blood" I inherited from my mother; on the other hand, many of my Black Italian collaborators and interlocutors in Italy are disenfranchised by the same law that allowed me to become an Italian citizen without having been born in Italy. I have felt out of place in Italy, attracting curious stares and looks of suspicion from white Italians; I have also seen ostensibly "white" Italian women in Naples with hair kinkier than mine and skin darker than my own. I have experienced white Italian cashiers express surprise when I speak to them in fluent Italian; I have also walked into bars in Milan's Eritrean Porta Venezia neighbourhood only to have a *barista* address me in Tigrinya (I am frequently mistaken in Italy for *Habesha*, or Eritrean/Ethiopian). I have witnessed some of my own *Lega*[5]-sympathetic relatives in Bergamo express a voyeuristic fascination with transnational Black popular culture in one breath, and in the next condemn the way that Italy has become "overrun" with African

[4] To disrupt the normative teleology of whiteness → Italianness → Italian citizenship, I make a point throughout this paper to mark what is typically left unmarked and refer to "white Italians" rather than simply "Italians."

[5] In the 2018 elections, the party formerly known as the *Lega Nord* (Northern League) dropped the *Nord* from its name in a bid to court voters from southern Italy—a constituency that had previously been one of the targets of the party's right-wing separatist rhetoric of *Padania*.

migrants in a way that threatens to pollute the country's cultural heritage. And when I lived in Milan, my elderly white Italian landlord often told me how much she sympathized with my parents' experiences of being in an interracial marriage because her own husband was a *meridionale*, from southern Italy.

As I have argued elsewhere (Hawthorne 2017; Hawthorne and Piccolo 2016), Italy's status as a Mediterranean crossroads has been framed since Italian national unification at the end of the nineteenth century alternatively as a *problem to be solved* (i.e., as a problem of degeneration, insufficient racial purity, or tenuous national coherence) or as a *potential solution to the challenges of modernity* (i.e., as a valuable source of cultural richness or economic interconnection). Italy's position at the southern edge of Europe and its large archaeological footprint of cross-Mediterranean civilizational encounter meant that it could not fit comfortably within the Aryanist raciologies that predominated in northern European countries such as France, Germany, and the United Kingdom in the nineteenth century (Gibson 1998; Haritaworn 2012).

In response, scientists, politicians, and policymakers working in the fledgling Italian nation were faced with the difficult task of fixing the blurry boundary lines between Europe, the Mediterranean, and Africa. These debates inspired an often-overlooked, homegrown school of Italian racial theorization whose most prominent figures included, among others, the internationally famous positive criminologist Cesare Lombroso (Caglioti 2017; Gibson 1998, 2002). The supposed racial ambiguity of Italians, the inheritance of thousands of years of trans-Mediterranean mixing, and Italy's geographical proximity to the African continent and the potential "contamination" of Blackness have ultimately worked to produce what John Gennari (writing in the related context of Italian American and African American cultural intersections) has evocatively characterized as a contact zone of "hopeful encounter and wary suspicion, dangerous, sometimes violent collision, and magnificent, joyous collusion" (Gennari 2017, p. 8).

As Angelica Pesarini argues, Italian citizenship law bears within it the traces of these contentious histories of struggle over the "true" racial identity of Italians (Pesarini 2017). Nonetheless, the regulation of Italian citizenship (and specifically, the emphasis on *jus sanguinis*, or right of blood) has been popularly regarded as a "natural" outgrowth of the Italian nation-building process and Italian mass emigration, or alternatively as a colour-blind policy that had unintentionally racial effects once Italy became a country of immigration characterized by increasingly visible demographic heterogeneity (Occhetta 2013). But the question of heterogeneity has actually shaped the shifting contours of Italian citizenship law since the formation of the Italian nation-state because of its aforementioned, liminal position in the Mediterranean—a promontory linking Europe with Africa (Njegosh 2015a, b).

Following Italian unification at the end of the nineteenth century, the dividing line of racial citizenship in Italy was primarily mapped onto the supposed anthropological and physiognomic distinction between northern and southern Italians (Giuliani and Lombardi-Diop 2013). But Italian colonialism in the Horn of Africa—which was unfolding alongside the *Risorgimento*—continued to raise intractable questions about the basis of Italian racial superiority and authority. Both the "racial proximity" of certain southern Italians to their colonized African subjects and the frequency of miscegenation in the colonies were increasingly viewed as threats to the dominant racial order (Andall and Duncan 2010; Giuliani 2014; Lombardi-Diop 2001). The Fascist period saw the gradual incorporation of southern Italians into the project of imperial expansion under the broad umbrella of "Mediterranean Aryanism" (De Donno 2006). By the late 1930s, however, this ideology eventually gave way to a—somewhat equivocal—embrace of Italian *Aryanism* (Gillette 2001, 2003; Giuliani 2014), and this shift was accompanied by restrictive racial laws targeting both Jews in Italy and Africans in the colonies.

This turn toward explicitly biological understandings of Aryan supremacy and the explicit socio-spatial regulations of interracial contact in late Fascist Italy are typically regarded as a sharp break from earlier understandings of racial difference and whiteness in Italy (a rupture often attributed to the influence of Nazi Germany). Yet even earlier Mediterraneanist theories of Italian racial identity stressed the persistence of an evolutionary gap between the people of the contemporary Italian peninsula and those of sub-Saharan Africa.[6] And this temporal convening of space (Massey 2005) ultimately produced a hierarchy of Mediterraneanness that was reflected in differential access to Italian citizenship. Even prior to the

[6]For instance, Giuseppe Sergi (a proponent of the "Mediterranean race" theory) argued that the Italian people originated in a Mediterranean race from the Ethiopian highlands. Still, he believed that in the present, there existed an evolutionary and civilizational divide between the branch of *homo eurafricus* that settled in Italy, and the contemporary people of the Horn of Africa (De Donno 2006; Fuller 2007; Giuliani 2014; Sergi 1895; Sòrgoni 2002).

infamous racial laws, for instance, the Italian government had already enacted a series of decrees and codes regulating the acquisition of citizenship by the "mixed-race" children of white Italian men and Black Eritrean women in the colonies. Mediterranean racial hierarchies were also reflected in Fascist-era citizenship laws, which distinguished between Libyan Italian citizens, Libyan Muslim citizens, and Italian *subjects*—a category reserved for Eritreans and Somalis (Bussotti 2016).

Italian legislation restricting citizenship for *meticci*, or children of white Italian fathers and Black Eritrean mothers, was not abolished until 1952 (Pesarini 2017). But even then, Italian citizenship was not "de-racialized." The current Italian citizenship law, Law no. 91, was enacted in 1992—a moment when concerns surrounding Italy's transition into a country of immigration, symbolized by the figure of the *African migrant*, began to feature prominently in political and popular discourse. This law reinforced the principle of *jus sanguinis* from the original 1912 Italian citizenship law: it simultaneously became easier for Italians in diaspora who had never set foot in Italy to re-acquire citizenship, and more difficult for immigrants and their Italian-born children to naturalize (*Acquisto della cittadinanza* 1992; Andall 2002; Zincone and Basili 2013). Law no. 91 also came directly on the heels of the 1990 Legge Martelli, Italy's first comprehensive immigration law, which formalized the immigration process by setting a yearly quota of migrants linked to the job market (Struggles in Italy, 2012). And while neither of these laws explicitly mentioned race nor directly targeted people of African descent, these policies cannot be understood separately from the longer trajectory of contestation about the basis of Italianness—contestation that has been framed by the poles of (white) Italian racial purity and Mediterranean mixedness.

While the more explicit racial nationalism of the Fascist era may have been at least temporarily laid to rest, contemporary Italian citizenship and immigration laws are in effect generating new forms of racialized differentiation based on supposedly race-neutral categories such as birthplace, descent, and neoliberal economic productivity (Bartoli 2012; Njegosh 2015a). In fact, one could argue that at a moment when explicit mentions of race are no longer publicly acceptable, *citizenship* has become a primary way to mediate the relationship of Blackness to Italianness. Legal nation-state citizenship by definition always draws distinctions between insiders and outsiders (Goldberg 2002; Isin 2005), distinctions that are subsequently treated as natural and inevitable. As Clelia Bartoli notes in her work on Italian institutional racism, for immigrants and their children "the condition of being a foreigner becomes like a biological fact from which it is impossible to escape" (Bartoli 2012, p. 64). In post-World War II Italy, biological understandings of *razza* [race] might have been thoroughly discredited, but *racism* and racialized disadvantage have clearly not vanished from the Italian sociopolitical and economic landscape (Lentin 2004; Mellino 2012; Romeo 2012). Instead, racism continues to reproduce and renew the category of race through new essentialisms that do not necessarily invoke blood or skin, but gain their meaning through other signifiers (Goldberg 2006; Hall 2000)—including citizenship.

The Twenty-First Century Mediterranean (Re)Turn

Given these intertwined histories of Italianness and Blackness, as well as the increasingly virulent backlashes against Afro-descendants in Italy, it is no surprise that the last decade and a half has witnessed a resurgence of academic and popular interest in the Mediterranean as a geographical unit of analysis. At a historical conjuncture marked by the rise of xenophobic, separatist, far-right political parties (Carter and Merrill 2007); the replacement of the Italian navy's Mare Nostrum search-and-rescue program with the EU's Triton maritime border securitization project (Stierl 2018); skyrocketing death tolls in the Mediterranean Sea (United Nations High Commissioner for Refugees 2017); the right-wing demonization of aid ships as covert migrant smuggling operations (Grenz et al. 2017); and progressively restrictive regulations of the acquisition of Italian citizenship (Massaro 2018), intellectuals and activists have sought to re-imagine the Mediterranean otherwise. This approach was expressed most provocatively by postcolonial theorist Iain Chambers in his 2008 text *Mediterranean Crossings*:

> Today, the patrolling of such borders and fears is focused and amplified in the boat people attempting the new middle passage across the Mediterranean. [...] The markers of race, religion, and racism frequently skew the question into a spiralling moral panic obsessively staring south toward Africa and Islam. Here a previous history—that of a multiple and mutable Mediterranean—is inadvertently evoked, only to be rapidly shut down and silenced. (Chambers 2008, pp. 9–10)

As Chambers goes on to argue, a critical re-appropriation of Mediterranean history requires a "*deconstruction* of being and becoming

Europe" (Chambers 2008, p. 10). In these sorts of retellings, the Mediterranean is understood as a space of fluid cartographics and hidden histories of contamination and creolization (e.g., *meticciato*) waiting to be exhumed from murky seafloor sediments. In other words, the Mediterranean represents a geographical and political counter-point to a global system of bounded, sovereign nation-states with restrictive citizenship. In this way, the Mediterranean as an analytical frame has the potential to throw into question the assumed isomorphism of European-Western-White and provide a powerful challenge to the rising tide of European ethnic absolutism (Gilroy 1993) in the twenty-first century.

Even beyond academia, since the turn of the twenty-first century the Italian publishing scene has witnessed a veritable explosion of texts addressing the question of meticciato and its relative degree of (in)compatibility with "Italo-Mediterranean" history and culture.[7] Mixing has become a powerful cipher encoding all sorts of debates about diversity, immigration, and the place of people of colour (and Black folk in particular) within the Italian nation. Perhaps most famously, former Minister of Integration Cécile Kyenge controversially declared in 2013 that "*L'Italia è meticcia*" ["Italy is a hybrid/mixed country"] during her advocacy in favour of *jus soli* citizenship for the children of immigrants who were born and raised in Italy ("Immigrazione, Kyenge: 'L'Italia è meticcia, lo ius soli sarà figlio del paese nuovo'" 2013). But what, if any, are the political limitations of this Mediterranean "hype of hybridity" (Broeck 2007; Mitchell 1997)—of invocations of Italy and the Mediterranean as a naturally transgressive third-space or zone of in-betweenness?[8] In the section that follows, I will turn to a event that revealed some of the drawbacks of using Mediterranean meticciato (as opposed to race and Blackness) as a point of reference in debates over who truly belongs in Italy.

[7] See https://books.google.com/ngrams for a Google N-Gram graph showing the frequency of Italian texts containing the word "meticciato" published between 1800 and 2000. Note that the other major spike in the occurrence of "meticciato" in Italian-language texts was during the Fascist period, when interracial mixing and the resulting the children of white Italian colonizers and Eritrean women were the subject of extensive state regulations.

[8] For a more general critique of the "betweenness of place," see Andrew Merrifield, "Place and Space: A Lefebvrian Reconciliation," *Transactions of the Institute of British Geographers* 18.4 (1993): 516–531. While debates about hybridity and in-betweenness sparked by the discursive turn in cultural studies (epitomized by the work of scholars such as Homi Bhabha) generally peaked around the end of the 1990s, these concepts still serve as influential frameworks for scholars and activists in Italy, given the history of racial formation in southern Europe.

Migrants, the Mediterranean, and the Burning Palms of Milan

At some point between the night of Saturday, 18 February, and the wee hours of Sunday, 19 February 2017, a group of young white Italian men bundled in heavy winter coats gathered under a cover of darkness to set fire to a cluster of palm trees that had recently been planted in Milan's Piazza del Duomo (D'Amico 2017). Their act of vandalism followed several days of highly-publicized outrage and racist polemic by politicians and everyday Italians about the symbolism of this new flora framing one of Milan's most iconic touristic spaces. In the press, the palm trees had been alternatively characterized as Miami-esque, exotic, tropical, and Mediterranean—all geographic descriptors that conjured vivid images of "mixed" people and cultures.

Vocal critics of the project included Matteo Salvini of the Lega, who declared that the palm trees were "Utter madness. The only things missing are sand and camels, and then the *clandestini*[9] will feel at home" ("Milano, bruciate tre palme in piazza Duomo" 2017). In a move reminiscent of the bananas thrown at Italy's first Black government minister Cécile Kyenge and Black Italian footballer Mario Balotelli (Christenson 2012; Zaccariello 2013), Lega protesters toted inflatable bananas and passed out real bananas to tourists in the piazza. (This was because there were also plans to plant bananas alongside the palm trees.) Nearby, members of the neo-fascist *Casa Pound* gathered to wave signs that read, "No to the Africanization of the Piazza del Duomo" ("Milano, bruciate tre palme in piazza Duomo" 2017). And just days prior to the arson, on 16 February 2017, a Facebook post by the group *Azione Identitaria* (which describes itself as a "patriotic movement of militant action and cultural struggle to protect Italic, regional, and European identity"[10]) issued a hyperbolic call to action that decried the palms and bananas as symbols of the ongoing destruction of Western civilization:

> According to these modern destroyers, not only should there no longer be any expressions of a man's land and his traditions, but they have to also destroy the symbols of a place, and every representation should demonstrate

[9] *Clandestini* is a term used by anti-immigration groups in Italy in reference to undocumented or irregular migrants.

[10] See "Chi Siamo," Azione Identitaria, https://www.azioneidentitaria.it/chi-siamo/ (accessed 9.7.2018).

> the end of specificity, of belonging, of the particularity of a population, of a nation and its symbols, including those of the natural world. (De Riccardis 2017)

The story of the burning palms in Milan evokes a dramatic story of intertwined spaces and regionalist racist backlashes against "transnational flows" at a moment marked by dramatic contestations over the symbolic and material borders of Italianness. From within the space of the nation-state, the Italian-born children of immigrants had been mobilizing for over a decade in an effort to reform Italy's restrictive *jus sanguinis* citizenship laws and facilitate the acquisition of Italian citizenship for almost one million disenfranchised youth of colour (De Franceschi 2018; "Immigrati, i numeri della seconda generazione" 2016).[11] At the same time, Italy—as one of the southernmost entry points into the European Union—was also embroiled in one of the largest refugee movements in recent history, as tens of thousands of asylum seekers from the African continent were fleeing on barely-seaworthy boats across the Mediterranean to escape violence, political instability, and economic hardship in their home countries. Thus, the palms and bananas came to stand in for all sorts of undesirable "African" contaminations of pristine (northern) Italian national space. And of course, the Duomo itself, as the largest cathedral in Italy,[12] was also a towering symbol of the enduring force of Catholicism as a "race-neutral" proxy for hegemonic understandings of the "unspoken" whiteness of Italian national identity (Giuliani 2014, p. 573; see also Portelli 2003, p. 29) (Fig. 9.1).

The palm tree controversy in many ways reflected the debates over the reform of Italian citizenship law that were unfolding simultaneously. The question of citizenship has re-emerged as one of the key terrains of struggle over the boundaries of race and nation in contemporary Italy, and this question is frequently framed as a referendum on Italy's relationship to the African continent. Many, if not most, of the publicly prominent *jus soli* activists are individuals of African descent. In addition, Black youth are

[11] The nationality law currently in force in Italy, Law no. 91 of 1992, specifies that the children of immigrants automatically inherit the nationality of their parents at birth. Upon reaching the age of majority (eighteen years old), they subsequently have a one-year window during which they can apply for Italian citizenship. To make this request, applicants must have documentary proof of continuous, legal residence in Italy from birth and pay an application fee of 200 Euros.

[12] St. Peter's Basilica is located in the state of Vatican City, not Italy.

Fig. 9.1 An "oasis" of palm and banana trees in the Piazza del Duomo of Milan. (Photo by Gigliola Hawthorne)

notably overrepresented in both campaign materials and media coverage of the citizenship reform movement—even though they actually do not make up the majority of the *seconda generazione* population in Italy (Istat 2016). On one particularly contentious day of debate in the Italian Senate about a citizenship reform bill during the summer of 2017, for instance, politicians from the Lega unfurled signs reading "*STOP INVASIONE*" [STOP THE INVASION] in bold letters ("Ius soli, discussione in Senato" 2017). The following day, a national right-wing newspaper printed the sensationalistic headline, "*CITTADINANZA FACILE: ITALIAFRICA*" [EASY CITIZENSHIP: ITALY-AFRICA] (Rame 2017).

The prospect of citizenship for the children of immigrants born in Italy has been conflated with neo-nationalist notions of "ethnic replacement," African invasion, and an ongoing demographic skirmish in which white Italian women must be incentivized to produce (white) Italian citizens. As a rather hyperbolic 2015 article published in the popular Italian weekly newsmagazine *L'Espresso* entitled "Italians in 50 Years? Hybrids or Disappeared" explained,

> By 2050, the Black Continent will have around a billion more people, compared to a Europe that risks losing 16 percent of this population. Today the population of young people aged 25 to 29—the class of emigrants by definition—is 51 million in Europe and 95 million in Africa. In only twenty years, there will be 41.12 million in Europe and a good 151 million in Africa; in thirty years, 40.9 million in Europe and 186 million in Africa. Their natural outlet, especially if Africa has not made the long-awaited leap forward in economic terms, will be the shores of the Mediterranean. "These are crazy numbers of adults with children," the demographers say, almost in unison: "It is time to think about what kind of society we want and begin taking steps in time". (Bianchi 2015)

At first glance, then, an impassioned defence of the trees in the Piazza del Duomo would seem to be an effective strategy for contesting the (quite literal) naturalization of the connection between race and nation (Daston and Vidal 2010; Moore et al. 2003; Somerville 2005). After all, as anthropologist Liisa Malkki argues, botanical metaphors of trees and roots work to territorialize peoples and cultures within bounded national spaces—positing migrants, refugees, and their children in turn as "unnatural," "pathologic," and "uprooted" subjects (Malkki 1992). And what better way to subvert arborescent models of Italian national belonging than with the family *Arecaceae*, whose native range includes the rim of the Mediterranean and stretches as far north as southern France? In this sense, then, a celebration of Milan's "Mediterranean" palms could flip these sedentarist and enclosed logics of place on their head, challenging the dangerous racial pretentions of northern Italian ethnic absolutism.

And indeed, following the burning of the palm trees, a self-proclaimed secular, pro–gay rights, anti-fascist, anti-racist group called *I Sentinelli di Milano* [The Guardians of Milan] organized a tongue-in-cheek protest through Facebook, urging residents of Milan to fill the Piazza del Duomo with "plants, fruits, and vegetables of all shapes, colours, and origins in æing Milan great" ("Palme bruciate a Milano, i Sentinelli sfidano Lega e CasaPound" 2017). The Sentinelli announcement on social media also sarcastically urged the "instigators of hate" to remove plants such as geraniums and irises from their balconies and window boxes, eliminate potatoes, tomatoes, and coffee from their diets, and discard their cell phones, since none of these things technically originated from or were manufactured within the boundaries of the Italian nation-state (I Sentinelli di Milano 2017). On the day of the counter-demonstration, a small group of mostly white Italian protesters descended upon the Piazza del Duomo carrying colourful images of the many fruits and vegetables that have become staples in Italian cuisine—tomatoes, eggplant, potatoes, corn, etc.—all emblazoned with their dates of "immigration" to Italy (e.g., zucchini, "*immigrata nel 1500*"). In other words, just as these plants had been *naturalized* as Italian, so too could African migrants and their children.

But upon closer inspection, the saga of the palm trees was not a simple story of Mediterranean hybridity triumphing over close-minded parochialism. Indeed, it was also a funhouse mirror of the racial politics of globalization and transnational neoliberal capitalism that have converged in contemporary Italy. The garden of forty-two palm trees and, eventually, fifty banana plants designed by Italian architect Marco Bay was actually funded by the Starbucks Corporation in preparation for the opening of its first Italian store in Italy—what would become the largest Starbucks coffeehouse in Europe (Henley 2018; "Milano, apre il più grande Starbucks d'Europa" 2018; Squires 2017). Even before the installation of the palm trees, the prospect of an American coffee chain entering the Italian market had been met with widespread scepticism and outright hostility (Maffei 2017). And Starbucks certainly did not intend to make a controversial political statement about African refugees, Italian national identity, or Italy's place in the Mediterranean through this particular dose of urban greenery. Rather, the plants were intended to add a dose of "*fin de siècle*" charm to the stately piazza—which had actually been flanked by palm trees back in the nineteenth century, around the time it took its present-day form (Squires 2017; ANSA 2017).

So, what are we to make of this bizarre conflation of neoliberal capitalism, multinational corporations, globalization, mass migration, and anti-Blackness? How did forty-two corporate palm trees come to stand in for—depending on who you ask—an unwanted African incursion in bounded Italian space, the homogenization of local particularity by the forces of transnational capitalism, Italy's exciting "new" diversity, or a beautiful and historic Mediterranean hybridity?

Racing Citizenship in the Black Mediterranean

The hype of meticciato in the Mediterranean Basin and beyond leaves us with many unanswered questions. Why are preoccupations about economic dispossession so frequently understood in terms that reify both racial and national essentialisms? Why are debates about *legitimate* vs. *illegitimate* or *deserving* vs. *undeserving* members of the national community typically framed in terms of economic productivity (Shilliam 2018)? And, in turn, why are so many celebratory narratives of transnationalism, mixing, and hybridity wrapped up with hegemonic narratives of transnational capitalism and neoliberal citizenship? After all, as Allan Pred wrote in the context of increasingly visible anti-Black racisms in Sweden in the late 1990s and early 2000s,

> Within mainstream neoliberal discourses…naturalizing and even glorifying "globalization," those feelings of exasperation, anxiety, and discontent, those feelings of disillusionment and betrayal, those feelings of actual or impending crisis, are instead frequently merged with other shallowly or deeply sedimented sentiments, frequently conjoined with unreflected collective assumptions regarding national culture and identity. […] Most prevalent of these…is the scapegoating of immigrants, refugees, and long-resident minorities, is the scapegoating of ethnic groups perceived as both social or cultural threats and as "illegitimate rivals in the struggle for scarce resources…". (Pred 2000, pp. 9–10)

The literal and metaphorical palm tree conflagration was but one snapshot in time. Nonetheless, reading the green tufts of the Starbucks palm trees against the neo-gothic spires of the Duomo di Milano can disclose a great deal about the questions of Mediterraneanism, hybridity, migration, and citizenship currently at stake in Italy today. After all, what the Lega and Casa Pound saw as evidence of racial and cultural degeneration, the Sentinelli in turn saw as a beautiful display of "hybrid vigour" (Daniel et al. 2014). Both the supporters and detractors of the palm trees hinged their arguments on different understandings of Italian hybridity—with far-right politicians blaming an encroaching cultural meticciato on multinational corporations and African migrants, and white activists from the liberal-left countering that "a *città meticcia* is pleasing" and that "A *Milano meticcia* is what we want" ("Palme e banani in piazza Duomo, ora 'spuntano' gli ortaggi. 'La diversità è un valore'" 2017).

But this discourse of Mediterranean mixing and conviviality, while seductive, ultimately proved inadequate because it elided the deep connections between racism, xenophobia, and dispossession in Italy. In other words, it functioned as a form of what David Theo Goldberg calls *racial neoliberalism*—a politics whereby racism and racist outcomes in the present are delinked from the conditions by which they were produced in the past (Goldberg 2009). By way of example, let us briefly return to the Sentinelli protest in Milan, during which counter-protesters carried pictures of "immigrant" fruits and vegetables. The implication was that just as these plants came to Italy from elsewhere, enriching Italian national culture, immigrants who arrive to Italy from across the Mediterranean today are also bringing with them colourful cultures, labouring bodies, and entrepreneurial spirits that can improve the city of Milan. This is a hybridity contingent upon migrants' conformation to the dominant local or national order, and to the tangible economic benefits brought by this cross-cultural contamination (Giglioli 2017a, pp. 15–16).

By centring questions of race and Blackness rather than abstract cosmopolitanisms, however, the Black Mediterranean responds to the racial neoliberalism of popular Mediterraneanisms in several important ways. When liberal-left activists and writers invoke the Mediterranean, it is usually in the context of Italy's supposed status as a *nazione meticcia*. In this way, the Mediterranean becomes an easy proxy for Italian race-neutrality—in other words, Italians do not have a racial identity because they are a "mixed" people, geographically proximate to Africa and not fully white, and are thus are incapable of racism. But rather than provide an opening for the inclusion of Black citizens within the body of the nation, these claims to Mediterranean racial innocence (Wekker 2016) inadvertently *reproduce* anti-Blackness in Italy. After all (as noted earlier), the Mediterranean has historically served as a polyvalent mechanism for drawing, revising, and contesting the boundaries of Italianness in relation to Africa and Blackness (De Donno 2006; Giuliani 2014). Even when Mediterraneanism was used as a fascist political strategy to legitimate Italian imperial expansion on the African continent, it still distinguished between the "Romanic" civilizations of North Africa and the more "racially backwards" populations of sub-Saharan Africa (Fuller 2007, p. 41; Giuliani 2014; Sòrgoni 2002).

In contemporary Italy, the incorporation of Black migrants and their families as potential citizens is contingent upon the extent to which they can blend seamlessly into a mixed Italian populace, shedding their difference entirely or reworking it into something that can be considered a "contribution" to the dominant national culture. As Goldberg explains, "Racial mixing may be desirable, but its product…is exhorted ultimately to mimic the cultural and performative standards of those embodying historical power" (Goldberg 2009, p. 344). Again, when the Sentinelli activists defended the palm trees in the Piazza del Duomo, they did so by effectively comparing Black migrants to vegetables—both being *objects* and *bodies* that have contributed in some way to the overall cultural and economic richness of Italy. A critical engagement with Italian complicity in the historical emergence of racial capitalism is thus supplanted with defanged stories of Mediterranean cosmopolitanism that uphold a vision of peaceful, economically-driven racial mixture in the present—or, to quote Goldberg's pithy summation, "Commerce thrives when people can interact and mix" (Goldberg 2009, p. 344).

The Black Mediterranean responds to these naturalizations by engaging with the non-deterministic ways in which the sedimentations and encrustations of the past have shaped the terrain of Black political action in the present (Danewid 2017). Cedric Robinson demonstrated in *Black Marxism*, for instance, that the Mediterranean Sea from the thirteenth through the sixteenth centuries was actually a laboratory for the forms of capital accumulation and labour exploitation that came to characterize the trans-Atlantic triangular trade (Robinson 2005; Saucier and Woods 2014). The merchants and financiers of the Italian maritime republics established extensive trade networks, employed enslaved labour, and provided capital and financing to support the Portuguese "voyages of discovery" in the Americas (Bono 2016; Robinson 2005). And of course, Italian colonialism in North Africa and the Horn of Africa, as well as its own internal colonization of the *Mezzogiorno* (Giglioli 2017b; Gramsci 2005; Schneider 1998; Verdicchio 1997), established a potent repertoire of racist practices, imagery, and modes of socio-spatial organization that continues to shape the lived experiences of Black communities in Italy to this day (Giuliani and Lombardi-Diop 2013; Merrill 2014; Pesarini 2017; Proglio 2016). This is why Saucier and Woods argue that any discussion of migration in the Italo-Mediterranean "that does not ground itself in the historical context of slavery and colonialism is imagining a world that is not, rather than dealing with the world as it is" (Saucier and Woods 2014, p. 60).

From Citizenship Struggles to Diasporic Ethics

As both the palm tree story and the persistent presence of "African invasion" in citizenship debates show, contestations over the boundaries of Italianness continue to be haunted the spectre of Blackness. This is why citizenship can become such a generative site for engaging with the complex racial politics of the Black Mediterranean. Sceptics of the citizenship reform movement on the left have argued, however, that activists are fundamentally misguided in thinking that they can trade a "racialized" notion of Italianness by blood (i.e., *jus sanguinis*) with a "race-neutral" Italianness based on birthplace and cultural fluency (i.e., *jus soli* and *jus culturae*[13]). In other words, they claim, these activists are falling victim to a naïve liberalism that places the hopes for *racial liberation* into the hands of the *racial state*. But blaming youth activists for the short-sightedness of focusing on the goal of citizenship commits yet another grave error of ahistoricism. Setting aside the fact that citizenship is materially consequential for the children of immigrants, casual dismissals of the citizenship reform movement also ignore the deeper histories of anti-Blackness in the Italo-Mediterranean region. These legacies have constrained political possibilities in the present, rendering citizenship a privileged terrain of struggle over racial inclusion in Italy. This is where the Black Mediterranean becomes useful not only as a theoretical framework, but as a model for *political praxis*. The Black Mediterranean allows us to connect the seemingly unbridgeable chasm between citizenship mobilizations and refugee rights struggles, situating them both in relation to a much longer history of contestation over the liminal racial identity of a Mediterranean nation whose own whiteness has always been precarious.

Devon Carbado's notion of "racial naturalization" is particularly useful in this regard. Racial naturalization refers to the ways in which the *experience of racism* effectively naturalizes a group into belonging to a nation,

[13] The most recent citizenship reform bill that was under consideration in the Italian parliament would have granted citizenship to the Italian-born children of immigrants on the condition that at least one parent was in possession of a long-term residency permit. The bill also introduced a form of citizenship acquisition called *jus culturae* [right of culture]. *Jus culturae* links citizenship for people who arrived to Italy as children to their time spent in the Italian school system. In other words, those who arrived to Italy before the age of 12 could become citizens after five years of school, while those who arrived between the ages of 12 and 18 could acquire citizenship after living in Italy for five years, completing a full scholastic cycle, and obtaining an educational qualification (e.g., a high school or vocational school diploma).

producing "inclusionary forms of exclusion" (Carbado 2005, p. 638). In other words, one becomes part of a nation through her or his insertion into particular racial hierarchies, regardless of whether that person has formal citizenship or not. Racial naturalization provides a model for understanding naturalization *outside* of the formal immigration process, capturing the wide variety of liminal social positions that exist at the intersection of formal citizenship, national identity, and equality (Carbado 2005, p. 642). The condition of being simultaneously inside and outside the nation thus becomes a way to connect seemingly disparate political struggles, foregrounding the racial underpinnings of the modern state rather than the liberal language of progressive inclusion, the racist teleology of "integration," or romantic invocations of national belonging. When we instead see *racism* as a naturalizing phenomenon that paradoxically produces groups as "Italian" or "Mediterranean," this perspective opens up new connections among Black subjects who are differentially positioned in relation to Italy's "racializing juridical assemblages" (Weheliye 2014, p. 79).

In a powerful 2018 article for the *Guardian*, the Italian-Somali writer, journalist, and activist Igiaba Scego responded to the frightening uptick in anti-Black racist violence in Italy with another hopeful invocation of Italy as the Mediterranean nation-state it could be:

> I have often wondered how it is that in such a young country, only 150 years old, there is such pronounced and deeply rooted racism. I have my own theory.
>
> In its heart, as a Mediterranean state, Italy knows itself to be a country with strong links to Africa. It could be the perfect pivot between continents, between Europe and Africa, yet it persists in denying its mixed-race identity as a country made of diversity. Everyone has passed through here: Arabs, Austrians, Africans, the French, the Spanish. This is Italy, a mixture of different blood and skins. When it finally accepts this identity, it will once again be the Bel Paese we all love. (Scego 2018)

While Scego's entreaty may at first glance appear to be yet another romantic invocation of Mediterranean conviviality in the spirit of the Sentinelli, it is actually so much more. For Scego, the Mediterranean is not predetermined as a space of peaceful, happy diversity—rather, she acknowledges that Italy's liminality is at the very heart of the current racist resurgence. After all, it is insufficient to respond to the rising tide of anti-Black racism and ethnonationalism in Italy by gesturing toward a suppressed history of mixing and intercultural contact. This move represents a form of political passivity, implying that the current insularity and xenophobia marring Italian politics is a deviation from a much longer history of trans-Mediterranean interconnection (in other words, meticciato was the past, so it must also be the future). Scego's Mediterraneanism, on the other hand, calls for *work*—the hard work of reckoning with Italy's colonial past and history of racial discrimination, and the labour of forging new solidarities that challenge the racist status quo. Scego's article represents one sliver of a broader vision of Black Mediterranean diasporic ethics: of anti-racist struggle oriented on shared trans-Mediterranean histories of dispossession rather than on naturalized notions of blood, culture, or territory.

What does this mean in terms of the struggle for citizenship reform in Italy today? The tempered *jus soli/jus culturae* bill approved by the lower house of the Italian parliament died in the Senate at the end of 2017 ("Ius soli, al Senato manca il numero legale. Assenti tutti i M5s e i centristi. Manca 1/3 dei Dem" 2017). And the 2018 *Decreto Salvini* threatens to further restrict access to Italian citizenship by increasing the waiting period for naturalization requests, raising the application fee, and introducing a new series of conditions under which citizenship can be revoked (Massaro 2018). But while many citizenship reform activists have become increasingly demoralized and hope to emigrate from Italy, this impasse also contains within it the seeds of other possibilities.

When the citizenship reform reached its peak between 2015 and 2017, the children of immigrants found themselves increasingly constrained to enact and embody Italianness in particular ways—sometimes quite literally, as in the case of the Italian-Senegalese designer Mbayeb "Mami" Bousso draping herself in an Italian flag gown to greet the President of Italy (Cartaldo 2017), and sometimes through acts of cultural performance, such as eating pasta or speaking in dialect. In other instances, activists attempted to re-appropriate histories of Mediterranean interconnection in the spirit of the *Sentinelli* by arguing that if Italy was once a great civilizational crossroads, the children of immigrants could resuscitate this suppressed cosmopolitanism through their embodied practices of cultural hybridity and transnational entrepreneurialism (Hawthorne 2018). Mobilizing in the midst of the refugee crisis, these activists sought to distinguish themselves (sometimes implicitly, and sometimes explicitly) in the public eye from newly arrived African refugees. As a result, however, they became inadvertently complicit in the process of drawing new racialized distinctions between "worthy" and "unworthy" potential Italian citizens, or "integratable" and "non-integratable" Afro-descendants.

It is important to remember, however, that these were never the only tactics used by the citizenship reform movement—though they may have garnered the most public attention. Earlier, *seconda generazione* activists had also worked to entirely de-familiarize the category of "Italian citizen" through media and the arts, pointing to complexities and wrinkles such as the Italian history of mass emigration (Clò 2012). And today, many Black activists are increasingly advocating for the need to disaggregate messy questions of identity from the bureaucratic matter of citizenship, approaching citizenship as a package of rights and capabilities[14] that can enable them to engage in other forms of outer-national politics. In other words, they are not rejecting the liberal politics of citizenship outright, but *swerving* them in creative new ways. As the work of Giuseppe Grimaldi (*this volume*) shows, for instance, Eritrean youth who were born and raised in Italy have engaged in a range of solidarity and *accoglienza* actions with newly arrived Eritrean refugees in Milan around the shared ties of diaspora and anti-colonial nationalism.

The wave of anti-Black violence in Italy spurred by the 2018 parliamentary elections (Momigliano 2018) has also pushed Afro-descendants in Italy into new alliances with each other that had seemed politically infeasible just years earlier. While the shootings in Macerata and Florence have been characterized in the press as manifestations of xenophobia and anti-immigrant sentiment (see, for instance, Riccardi 2018), even Black Italians who were born in Italy increasingly find themselves at risk on Italian streets because—as A. Sivanandan famously wrote—they all wear their passports on their faces ("Racism, liberty and the war on terror" 2007, p. 48). And in southern Italy, new and unexpected coalitions are emerging among groups that share histories of marginalization as "southern"—from Black youth born and raised in Italy, to African migrants and refugees working

[14] The children of immigrants who were born in Italy but lack Italian citizenship cannot go on school-sponsored field trips outside of Italy without first requesting a visa. They cannot apply for jobs through the state system of *concorsi pubblici*. They cannot vote in local or national elections. Those who are unable to successfully petition for Italian citizenship upon turning eighteen must live in Italy on the equivalent of a long-term residency permit (which requires, among other stipulations, the maintenance of a minimum income level). The loss of a residency permit bears with it the possibility of deportation to a parental home country they might never have visited.

in exploitative agricultural labour camps, to (white) southern Italians living precariously in an historically racialized region of the peninsula ("I Terroni Uniti le cantano a Salvini: 30 artisti contro il razzismo in 'Gente do sud'" 2017; Terroni Uniti 2017). Rather than flattening diverse experiences of dispossession in order to minimize the gravity of anti-Black racism in contemporary Italy, these groups are instead drawing on historical parallels to activate new kinds of alliances. Taken together, these various examples constitute both large and small subversions of the Italian state's legal-racial categories, and produce alternative imaginaries of community and solidarity. In the spirit of W.E.B. Du Bois (Du Bois 2011; Slater 2014), they form a capacious diasporic ethics based not on skin colour or legal status, but on the shared social kinship of being naturalized into the racial regimes of Italy and the Mediterranean.

Afterword: The Relevance of the Black Mediterranean in the Time of Monsters[15]

These multifaceted Black Mediterranean interventions and mobilizations represent insurgent practices of Black diasporic knowledge production. There is, after all, no institutional structure for Black studies in Italy, and because of long-standing attempts to externalize Blackness as something foreign to the Italian nation (Lombardi-Diop 2011), efforts to bring race-critical theory into Italy are often rejected as Anglo-American impositions.[16] The Black Mediterranean provides an analytical vocabulary for engaging with the historically- and geographically-specific production of racisms in Italy in a way that is also attentive to the transnational circuits of ideas, technologies of power, and modes of resistance. But what is the broader significance of the Black Mediterranean—as both theory and practice—beyond the shores of the Italian peninsula?

[15] The "time of monsters" is derived from the quote by Antonio Gramsci, "*La crisi consiste appunto nel fatto che il vecchio muore e il nuovo non può nascere: in questo interregno si verificano i fenomeni morbosi più svariati*" [The crisis consists precisely in the fact that the old is dying and the new cannot be born; in this interregnum a wide variety of morbid phenomena appear" (Gramsci 2001, p. 311). This excerpt from the Prison Notebooks is commonly styled into English as "The old world is dying and the new world struggles to be born. Now is the time of monsters" (Gramsci 1971, p. 276).

[16] Important efforts to engage race critical theory from within the Italian academy include the research networks InteRGRace, Postcolonialitalia, and Slanting Gaze on Social Control, Labour, Racism, and Migration.

Italy is, in many regards, a canary in the coalmine for the global resurgence of the far right—as seen in Matteo Salvini's feature on the cover of the United States–based *Time* magazine in 2018 (Walt 2018). While Cedric Robinson might have viewed the Black Mediterranean as a *precondition* for Black Atlantic regimes of racial capitalism, today the ongoing *reproduction* of the Black Mediterranean provides a window onto the interlocking processes of economic dispossession, postcolonialism, xenophobia, and racism shaping the modern world. In this way, Black communities in Italy can longer be reduced to passive receptacles for a linear diffusion of diasporic resources out of North America. Instead, they are generating creative new political practices and imaginaries that can inform struggles on the other side of the Atlantic as well.

References

Acquisto della cittadinanza. 1992. Legge 5 febbraio 1992, n. 91 – Nuove norme sulla cittadinanza. Ministero dell'Interno, Circolare 11 novembre 1992, n. 601 (GU Serie Generale n.279 del 26-11-1992), https://www.gazzettaufficiale.it/eli/id/1992/11/26/092A5478/sg

Affricot, J. 2015. Lidia Carew "Sono un'ITALIANA con la pelle nera, NON una nera che si SENTE italiana." GRIOT. http://griotmag.com/it/lidia-carew-un-italiana-la-pelle-nera-non-nera-si-sente-italiana/. Accessed 17 Sept 2018.

Agamben, G. 1998. *Homo Sacer: Sovereign Power and Bare Life*. Stanford: Stanford University Press.

———. 2017. Perché non ho firmato l'appello sullo ius soli. Quodlibet, October 18, 2017. https://www.quodlibet.it/giorgio-agamben-perch-onho-firmato-l-appello-sullo-ius-soli.

Andall, J. 2002. Second-Generation Attitude? African-Italians in Milan. *Journal of Ethnic and Migration Studies* 28: 389–407. https://doi.org/10.1080/13691830220146518.

Andall, J., and D. Duncan. 2010. *National Belongings: Hybridity in Italian Colonial and Postcolonial Cultures*. Bern: Peter Lang.

ANSA. 2017. Starbucks Plants Palm Trees in Milan's Piazza Duomo, February 16, 2017. http://www.ansa.it/english/news/2017/02/16/starbucks-plants-palmtrees-in-milans-piazza-duomo_be969f37-9e0d-4350-b950-bb4314d78ea1.html.

Bartoli, C. 2012. *Razzisti per legge: L'Italia che discrimina*. Rome: Gius. Laterza & Figli Spa.

Bianchi, F. 2015. Gli italiani fra 50 anni? O meticci o scomparsi [WWW Document]. l'Espresso. http://espresso.repubblica.it/attualita/2015/09/24/news/gli-italiani-fra-50-anni-o-meticci-o-scomparsi-1.231363. Accessed 27 Sept 2018.

Bono, S. 2016. *Schiavi: una storia mediterranea (XVI-XIX secolo)*. Bologna: Il Mulino.

Broeck, S. 2007. White Fatigue, or, Supplementary Notes on Hybridity. In *Reconstructing Hybridity: Postcolonial Studies in Transition*, ed. J. Kuortti and J. Nyman, 43–58. Amsterdam: Rodopi.

Bussotti, L. 2016. A History of Italian Citizenship Laws during the Era of the Monarchy (1861–1946). *Advances in Historical Studies* 05: 143. https://doi.org/10.4236/ahs.2016.54014.

Caglioti, A.M. 2017. Race, Statistics and Italian Eugenics: Alfredo Niceforo's Trajectory from Lombroso to Fascism (1876–1960). *European History Quarterly* 47: 461–489. https://doi.org/10.1177/0265691417707164.

Carbado, D.W. 2005. Racial Naturalization. *American Quarterly* 57: 633–658.

Cartaldo, C. 2017. Immigrata veste col Tricolore: denunciata per vilipendio. *Il G.*

Carter, D., and H. Merrill. 2007. Bordering Humanism: Life and Death on the Margins of Europe. *Geopolitics* 12: 248–264. https://doi.org/10.1080/14650040601168867.

Chambers, I. 2008. *Mediterranean Crossings: The Politics of an Interrupted Modernity*. Durham: Duke University Press.

Christenson, M. 2012. Euro 2012: Mario Balotelli Threatens to "Kill" Banana-Throwing Fans. *The Guardian*.

Clò, C. 2012. Hip Pop Italian Style: The Postcolonial Imagination of Second-Generation Authors in Italy. In *Postcolonial Italy: Challenging National Homogeneity*, ed. C. Lombardi-Diop and C. Romeo, 275–292. New York: Palgrave Macmillan.

D'Amico, P. 2017. Palme bruciate a Milano: un ragazzo con il montgomery. *Corriere Della Sera*.

Danewid, I. 2017. White Innocence in the Black Mediterranean: Hospitality and the Erasure of History. *Third World Quarterly* 38: 1674–1689. https://doi.org/10.1080/01436597.2017.1331123.

Daniel, G.R., L. Kina, W.M. Dariotis, and C. Fojas. 2014. Emerging Paradigms in Critical Mixed Race Studies. *Journal of Critical Mixed Race Studies* 1: 40.

Daston, L., and F. Vidal. 2010. Doing What Comes Naturally. In *The Moral Authority of Nature*, ed. L. Daston and F. Vidal, 1–19. Chicago: University of Chicago Press.

De Donno, F. 2006. La Razza Ario-Mediterranea. *Interventions* 8: 394–412. https://doi.org/10.1080/13698010600955958.

De Franceschi, L. 2018. *La cittadinanza come luogo di lotta. Le seconde generazioni in Italia fra cinema e serialità*. Rome: Arcane Editrice.

De Riccardis, S. 2017. Palme bruciate a Milano, c'è l'identikit del vandalo. La polizia locale: "Si presenti e chieda scusa". *La Repubblica*.

Di Maio, A. 2012. Mediterraneo nero. Le rotte dei migranti nel millennio globale. In *La Citta' Cosmopolita*, ed. G. de Spuches and V. Guarrasi, 142–163. Palermo: Palumbo Editore.

Du Bois, W.E.B. 2011. *Dusk of Dawn: An Essay Toward an Autobiography of a Race Concept*. New Brunswick: Transaction Publishers.

Foucault, M. 1990. *The History of Sexuality, Vol. 1: An Introduction*. New York: Vintage.

Fuller, M. 2007. *Moderns Abroad: Architecture, Cities and Italian Imperialism*. London: Routledge.

Gennari, J. 2017. *Flavor and Soul: Italian America at Its African American Edge*. Chicago: University of Chicago Press.

Gibson, M. 1998. Biology or Climate? Race and "Deviance" in Italian Criminology, 1880–1920. In *Italy's "Southern Question": Orientalism in One Country*, ed. J. Schneider, 99–116. Oxford: Berg.

———. 2002. *Born to Crime: Cesare Lombroso and the Origins of Biological Criminology*. Westport: Praeger.

Giglioli, I. 2017a. From "a Frontier Land" to "a Piece of North Africa in Italy": The Changing Politics of "Tunisianness" in Mazara del Vallo, Sicily. *International Journal of Urban and Regional Research* 41: 749–766. https://doi.org/10.1111/1468-2427.12544.

———. 2017b. Producing Sicily as Europe: Migration, Colonialism and the Making of the Mediterranean Border Between Italy and Tunisia. *Geopolitics* 22: 407–428. https://doi.org/10.1080/14650045.2016.1233529.

Gillette, A. 2001. The Origins of the "Manifesto of Racial Scientists". *Journal of Modern Italian Studies* 6: 305–323. https://doi.org/10.1080/13545710110084253.

———. 2003. *Racial Theories in Fascist Italy*. London: Routledge.

Gilroy, P. 1993. *The Black Atlantic: Modernity and Double Consciousness*. Cambridge, MA: Harvard University Press.

Giuliani, G. 2014. L'Italiano Negro. *Interventions* 16: 572–587. https://doi.org/10.1080/1369801X.2013.851828.

Giuliani, G., and C. Lombardi-Diop. 2013. *Bianco e nero. Storia dell'identità razziale degli italiani*. Milan: Mondadori Education.

Gnisci, A. 1998. *Creoli, meticci, migranti, clandestini e ribelli*. Rome: Meltemi Editore srl.

Goldberg, D.T. 2002. *The Racial State*. Malden: Blackwell Publishers.

———. 2006. Racial Europeanization. *Ethnic and Racial Studies* 29: 331–364. https://doi.org/10.1080/01419870500465611.

———. 2009. *The Threat of Race: Reflections on Racial Neoliberalism*. Malden: Wiley-Blackwell.

Gramsci, A. 1971. *Selections from the Prison Notebooks*. New York: International Publishers.

———. 2001. *Quaderni dal carcere*. Einaudi: Turin.

———. 2005. *The Southern Question*. Toronto: Guernica Editions.

Grenz, C., S. Slan, and J. Lier. 2017. Developments in the Central Mediterranean Over the Past Two Years. Watch Med Alarm Phone. https://alarmphone.org/

en/2017/10/18/developments-in-the-central-mediterranean-over-the-past-two-years/. Accessed 12 Sept 2018.

Hall, S. 2000. Conclusion: The Multi-cultural Question. In *Un/Settled Multiculturalisms: Diasporas, Entanglements, Transruptios*, ed. B. Hesse, 209–241. New York: St. Martin's Press.

Haritaworn, J. 2012. *The Biopolitics of Mixing: Thai Multiracialities and Haunted Ascendancies*. London: Routledge.

Harris, C.I. 1993. Whiteness as Property. *Harvard Law Review* 106: 1707–1791. https://doi.org/10.2307/1341787.

Hawthorne, C. 2017. In Search of Black Italia: Notes on Race, Belonging, and Activism in the Black Mediterranean. *Transition* 123: 152–174.

———. 2018. *There Are No Black Italians? Race and Citizenship in the Black Mediterranean*. Berkeley: UC Berkeley.

Hawthorne, C., and P. Piccolo. 2016. *"Meticciato" o della problematicità di una parola*. Macch: Sognante.

Henley, J. 2018. Grande Opening: Milan Gets Italy's First Starbucks. *The Guardian*.

I Sentinelli di Milano. 2017. Piantatela lÌ. Coltiviamo la diversità [WWW Document]. Facebook. https://www.facebook.com/events/1330561903671695/. Accessed 7 Sept 2018.

I Terroni Uniti le cantano a Salvini: 30 artisti contro il razzismo in "Gente do sud". 2017. *Adnkornos*.

Immigrati, i numeri della seconda generazione. 2016. *La Repubblica*.

Immigrazione, Kyenge: "L'Italia è meticcia, lo ius soli sarà figlio del paese nuovo". 2013. *Il Fatto Quotid*.

Isin, E.F. 2005. Engaging, Being, Political. *Political Geography* 24: 373–387.

Istat. 2016. Cittadini non comunitari: presenza, nuovi ingressi e acquisizioni di cittadinanza [WWW Document]. http://www.istat.it. Accessed 28 Mar 2018.

Ius soli, al Senato manca il numero legale. Assenti tutti i M5s e i centristi. Manca 1/3 dei Dem. 2017. *La Repubblica*.

Ius soli, discussione in Senato: è bagarre. Contusa ministra Fedeli e Boldrini dice basta alla violenza in Aula. 2017. *La Repubblica*.

Lentin, A. 2004. *Racism and Anti-Racism in Europe*. London: Pluto Press.

Lombardi-Diop, C. 2001. Mothering the Nation: An Italian Woman in Colonial Eritrea. In *ItaliAfrica: Bridging Continents and Cultures*, ed. S. Matteo. Stony Brook: Forum Italicum.

———. 2011. Spotless Italy: Hygiene, Domesticity, and the Ubiquity of Whiteness in Fascist and Postwar Consumer Culture. *California Italian Studies* 2.

Maffei, L. 2017. Starbucks to Open in Italy, Home of Espresso, in 2018. Italian Cafes Say Bring It. *NPR*.

Malkki, L. 1992. National Geographic: The Rooting of Peoples and the Territorialization of National Identity Among Scholars and Refugees. *Cultural Anthropology* 7: 24–44.

Massaro, A. 2018. Cittadinanza italiana da 730 giorni a 4 anni, tutte le novità del decreto salvini sulle cittadinanze. Cittadinanza Ital.

Massey, D.B. 2005. *For space*. London/Thousand Oaks: Sage.

Mellino, M. 2012. De-Provincializing Italy: Notes on Race, Racialization, and Italy's Coloniality. In *Postcolonial Italy: Challenging National Homogeneity*, ed. C. Lombardi-Diop and C. Romeo, 83–99. New York: Palgrave Macmillan.

Merrill, H. 2014. Postcolonial Borderlands: Black Life Worlds and Relational Place in Turin, Italy. *ACME: An International Journal for Critical Geographies* 13: 263–294.

———. 2018. *Black Spaces: African Diaspora in Italy*. New York: Routledge.

Milano, apre il più grande Starbucks d'Europa: ecco com'è dentros. 2018. *Repubblica Tv*.

Milano, bruciate tre palme in piazza Duomo. 2017. *La Stampa*.

Mitchell, K. 1997. Different Diasporas and the Hype of Hybridity. *Environment and Planning D: Society and Space* 15: 533–553. https://doi.org/10.1068/d150533.

Momigliano, A. 2018. With Anti-Immigrant Feeling on the Rise, Italy's African Community Shocked by Another Killing. *The Washington Post*.

Moore, D.S., A. Pandian, and J. Kosek. 2003. The Cultural Politics of Race and Nature. In *Race, Nature, and the Politics of Difference*, ed. D.S. Moore, J. Kosek, and A. Pandian, 1–70. Durham: Duke University Press.

Njegosh, T.P. 2015a. La finizione della razza, la linea del colore e il meticciato. In *Il colore della nazione*, ed. G. Giuliani, 213–227. Milan: Mondadori.

———. 2015b. Il Meticciato Nell'Italia Contemporanea. Storia, Memorie e Cultura di Massa. Iperstoria—Testi Lett. *Linguaggi* 6: 143–166.

Occhetta, F. 2013. La cittadinanza in Italia. *Civiltà Cattolica* IV: 14–24.

Palme bruciate a Milano, i Sentinelli sfidano Lega e CasaPound: "In piazza con piante da tutto il mondo. 2017. *La Repubblica*.

Palme e banani in piazza Duomo, ora "spuntano" gli ortaggi. 'La diversità è un valore'. 2017. *Il Giorno*.

Pesarini, A. 2017. "Blood is Thicker than Water": The Materialization of the Racial Body in Fascist East Africa. *Zapruder World: An International Journal for the History of Social Conflict* 4. https://doi.org/10.21431/Z33S32.

Portelli, A. 2003. The Problem of the Color Blind: Notes on the Discourse of Race in Italy. In *Crossroutes—The Meaning of Race for the 21st Century*, ed. P. Boi and S. Broeck, 29–39. Hamburg: LIT.

Pred, A. 2000. *Even in Sweden: Racisms, Racialized Spaces, and the Popular Geographical Imagination*. Berkeley: University of California Press.

Proglio, G., ed. 2016. *Decolonizing the Mediterranean*. Cambridge: Cambridge Scholars Publishing.

———. 2018. Is the Mediterranean a White Italian–European Sea? The Multiplication of Borders in the Production of Historical Subjectivity. *Interventions*: 1–22. https://doi.org/10.1080/1369801X.2017.1421025.

Racism, Liberty and the War on Terror. 2007. Race Cl. 48, 45–96. https://doi.org/10.1177/0306396807077071

Raeymaekers, T. 2015a. Working the Black Mediterranean. *Liminal Geogr.* http://www.timothyraeymaekers.net/2015/01/working-the-black-mediterranean/. Accessed 10 Apr 2017.

———. 2015b. The Racial Geography of the Black Mediterranean. *Liminal Geogr.* http://www.timothyraeymaekers.net/2015/01/the-racial-geography-of-the-black-mediterranean/. Accessed 10 Apr 2017.

Rame, S. 2017. Ius soli, è sostituzione etnica: subito 800mila nuovi "italiani". *Il G.*

Riccardi, K. 2018. Roberto Saviano sul Guardian: "Il fascismo è tornato, e sta paralizzando l'Italia". *La Repubblica.*

Robinson, C.J. 2005. *Black Marxism: The Making of the Black Radical Tradition.* Chapel Hill: University of North Carolina Press.

Romeo, C. 2012. Racial Evaporations: Representing Blackness in African Italian Postcolonial Literature. In *Postcolonial Italy: Challenging National Homogeneity*, ed. C. Lombardi-Diop and C. Romeo, 221–236. New York: Palgrave Macmillan.

Saucier, P.K., and T.P. Woods. 2014. Ex Aqua: The Mediterranean Basin, Africans on the Move, and the Politics of Policing. *Theoria* 61: 55–75. https://doi.org/10.3167/th.2014.6114104.

Scego, I. 2018. Italy Is My Country—But It Must Face Its Racist History. *The Guardian.*

Schneider, J. 1998. *Italy's "Southern Question": Orientalism in One Country.* New York: Bloomsbury Academic.

Sergi, G. 1895. *Origine e diffusione della stirpe mediterranea.* Rome: Società editrice Dante Alighieri.

Shilliam, R. 2018. *Race and the Undeserving Poor: From Abolition to Brexit.* Newcastle upon Tyne: Agenda Publishing.

Slater, A.M. 2014. W.E.B. Du Bois' Transationalism: Building a Collective Identity Among the American Negro and the Asian Indian. *Phylon* 1960–51: 145–157.

Somerville, S.B. 2005. Notes Toward a Queer History of Naturalization. *American Quarterly* 57: 659–675.

Sòrgoni, B. 2002. Racist Discourses and Practices in the Italian Empire Under Fascism. In *The Politics of Recognizing Difference: Multiculturalism Italian Style*, ed. R. Grillo and J. Pratt, 41–57. Aldershot: Ashgate.

Squires, N. 2017. Locals Deride "Kitsch" Starbucks-Sponsored Palm Trees Planted in Milan's Cathedral Square. *The Telegraph.*

Stierl, M. 2018. A Fleet of Mediterranean Border Humanitarians. *Antipode* 50: 704–724. https://doi.org/10.1111/anti.12320.

Struggles in Italy. 2012. Immigration Policies in Italy, July 11, 2012. https://strugglesinitaly.wordpress.com/equality/enimmigration-policies-in-italy/.

Terroni Uniti. 2017. Gente Do Sud. https://www.youtube.com/watch?v=TVKGGyoUlRo.

Thompson, R.F. 2010. *Flash of the Spirit: African & Afro-American Art & Philosophy*. New York: Random House.

United Nations High Commissioner for Refugees. 2017. Italy: UNHCR Update #10. https://data2.unhcr.org/en/documents/details/53633.

Verdicchio, P. 1997. The Preclusion of Postcolonial Discourse in Southern Italy. In *Revisioning Italy: National Identity and Global Culture*, ed. B. Allen and M.J. Russo, 191–212. Minneapolis: University of Minnesota Press.

Walt, V. 2018. Why Italy's Matteo Salvini Is the Most Feared Man in Europe. *Time*.

Weheliye, A.G. 2014. *Habeas Viscus: Racializing Assemblages, Biopolitics, and Black Feminist Theories of the Human*. Durham: Duke University Press.

Wekker, G. 2016. *White Innocence: Paradoxes of Colonialism and Race*. Durham: Duke University Press.

Zaccariello, G. 2013. Kyenge, banane contro il ministro alla festa Pd. Lei: "Razzismo sintomo di disagio". *Il Fatto Quotid*.

Zincone, G., and M. Basili. 2013. *Country Report: Italy, EUDO Citizenship Observatory*. Florence: European University Institute.

CHAPTER 10

Reimagining Citizenship in the Black Mediterranean: From *Jus Sanguinis* to *Jus Soli* in Contemporary Italy?

Vivian Gerrand

The author wishes to acknowledge the Australian European University Institute Fellowships Association Inc. for the award of a six-month Postdoctoral Fellowship at the European University Institute in 2015 that made it possible to pursue the research project: 'From *jus sanguinis* to *jus soli*: Representing Somalis' Multiple Affiliations.' I am grateful to Professor Anna Triandafyllidou, director of the Cultural Pluralism area of the Global Governance Programme at the Robert Schuman Centre for Advanced Studies, for facilitating the Migration Working Group *Reimmaginare la cittadinanza* [Re-imagining Citizenship] encounter on which this chapter is based. Special thanks to Kaha Mohamed Aden, Marilena Fabbri, Cadigia Hassan, Fred Kuwornu, Antar Mohamed Marincola and Antonio Morone for participating in the workshop and to Asha Sabrie for filming the event.

V. Gerrand (✉)
Alfred Deakin Institute for Citizenship and Globalisation, Deakin University, Melbourne, VIC, Australia
e-mail: vivian.gerrand@deakin.edu.au

G. Proglio et al. (eds.), *The Black Mediterranean*, Mediterranean Perspectives, https://doi.org/10.1007/978-3-030-51391-7_10

> When I started fighting for the right to Italian citizenship for the children of migrants I was twenty and I had really black hair. Now I'm forty, with some grey streaks, and nothing has changed. We're still viewed as foreigners… strangers in our own country.[1] (Igiaba Scego 2017)

State Citizenship in the Twenty-First Century

Citizenship has been conceived of as the formal instrument through which an immigrant acquires equal rights and obligations akin to those of the native population. It may be a milestone and a tool for a migrant's integration or it may be conceptualized as the end of the integration journey. However, it can become a challenge to democracy when access to citizenship is severely restricted (*jus soli, jus domicili*) and when it is provided only to those with co-ethnic ties with the nation (*jus sanguinis*). Laws surrounding who can be a citizen of a state are rarely uncontested and may become sites of conflict, especially in the twenty-first century's heightened levels of human global mobility, because they establish the boundaries of national belonging (Colombo et al. 2011; Tintori 2018).

To illuminate the relationship between citizenship and the hidden and silenced histories of the Black Mediterranean, in this chapter I take Italy's citizenship law and its proposed reforms within the contemporary cultural context as a case study of racialized citizenship (Pesarini 2017; Proglio 2018; Sardelic 2019; Scego 2018). Italy's family-oriented *jus sanguinis* citizenship laws exclude approximately 8.1% of its resident population (Istat 2017). The lack of political representation for such a significant part of the population produces an "evident detachment of the Italian society from its political community" (Zincone and Basili 2013). Indeed, in a profound departure from the basic norms of democratic legitimisation of political authority, resident 'foreign' citizens in Italy bear many of the community's burdens without receiving its benefits.

Conceptualising Citizenship in the Black Mediterranean

The mismatch between those who live and feel Italian and experience 'thick' cultural citizenship and those who can access 'thin' official citizenship is not without historical precedent (Bauböck 1999; Gerrand 2016a,

[1] Unless otherwise specified, all translations from Italian into English in this chapter are my own.

b; Stevenson 2003). Informed by the racialized partitioning of Italy's colonial subjects in the Horn of Africa, the 1992 Law 91 was designed to exclude immigrants while including Italian emigrants (Morone 2015, 2018). Accordingly, the everyday lived realities of being an immigrant Italian are still not substantially acknowledged within its political systems, nor are they accurately represented in dominant Italian media, even after decades of immigration being a structural reality and critical pillar of the Italian economy (Interview with Gerrand 2019; Gillio 2017; Giustiniani 2003; Nadotti 2018; Queirolo Palmas 2006).

Since 2009, it has been a crime to be an unauthorised migrant in Italy. A further decree introduced in October 2018 by the Di Maio Salvini administration banned humanitarian residency in the country, provided extra funding to deportation centres and made it possible to strip citizenship from Italian citizens suspected of crimes if they have a migrant background (Camilli 2018). The decree also led to evacuations of welcome centres, which left vulnerable migrants on the streets in a number of cities (Tondo and Giuffrida 2018).

In this climate of closure, it may be hard to countenance that in 2012, when then President of the Italian Republic Giorgio Napolitano decried as folly[2] (Moual 2012) the reluctance to grant the children of immigrants in Italy Italian citizenship, public opinion was overwhelmingly in favour (72%) of granting citizenship to Italian-born children (Istat 2017). Yet in spite of cohesive citizenship law reform campaigns with strong media and political support (including from the Catholic Church) and the laws having been approved in the lower house, they never made it into legislation and were shelved in 2017 (Interview with Fabbri 2019; Tintori 2018). Since that time, the rise of the vocal far right's "retrotopian" and populist rhetoric (Bauman 2017; Goodhart 2017) has posed a growing threat to Italy's cultural, ethnic and religious diversity. Its increasingly securitised (but not secure) mechanisms of governance cast doubt on the future prospects of Italian democracy, without which the proposed citizenship amendments would unlikely ever be implemented.

[2] ["L'ostilità nei confronti del l'immigrazione... deve essere considerata un rifiuto della realtà, frutto di ingiustificate paure troppo spesso alimentate nel dibattito pubblico. Il fenomeno immigratorio, che è inevitabile, deve essere perciò accompagnato da politiche adeguate, perché a coloro che vengono a lavorare in Italia sia attribuito il rispetto che meritano, nell'osservanza delle nostre leggi."] https://www.ilsole24ore.com/art/notizie/2012-12-19/napolitano-cittadinanza-figli-immigrati-064057.shtml?refresh_ce=1

Chapter Outline

To investigate contemporary mobilisations of racialized citizenship in the Mediterranean, the first section of the chapter considers who is regarded as Italian today and how dominant conceptions of what it means to be Italian are upheld through exclusivist narratives conveyed via images. Narratives fostering social division and undermining inclusive belonging and citizenship in Italy are not new. What is new are the specific affordances created by digital media networks through multimodal communication (Kress 2009; O'Halloran and Smith 2011); Papacharissi 2010), which bring text, image, sound and kinesthetics together in novel ways that deeply engage sensory as well as cognitive trajectories of experience, feeling and action. If we accept that, to a significant degree, communities are established via images and representations, then these are able to 'do' and 'undo' projects of conviviality. Thus, to deepen understanding of how race is operationalised to partition people born in contemporary Italy through citizenship, I undertake a multimodal analysis of a selection of viral online images circulated in the past three years.

In dialogue with the *Black Mediterranean* project's reclaiming of histories that have been silenced, oppressed, and marginalised, the second section of this chapter reflects upon the 'Reimagining Citizenship' encounter that was organized by the author in collaboration with the Migration Working Group at the European University Institute in April 2015. With the aim of investigating citizenship and national identity in Italy today, paying special attention towards the predicament of migrants in Italy from the country's former colonies, the 'Reimagining Citizenship' event provided a platform for Italian intellectuals of Somali background Cadigia Hassan, Antar Marincola and Kaha Mohamed Aden to discuss their writings on what it means to be an Italian with Somali heritage. Filmmaker Fred Kuwornu also attended the event and screened his documentary *18 Ius Soli*. A discussion of proposed Italian citizenship law reform then took place between Italian Member for Parliament Marilena Fabbri, Dr. Antonio Morone and Professor Anna Triandafyllidou.

This section will focus on the ways in which different engagements with citizenship—be they literary, political or historical—have important roles to play in shaping citizenship law reform. An in-depth recent interview with former MP Marilena Fabbri of the then-governing Democratic Party will offer a point of entry to exploring the extent to which such representations are in dialogue with proposed reforms to Italian citizenship policy.

In many cases, this has involved representing and thus giving a political platform to the lived and imagined realities of Italy's colonial past and the ways in which this history continues to shape the country's present. It also enables engagement with thinking that enlarges the project of citizenship acquisition beyond the borders of any given nation-state (Balibar 2004; Finotelli et al. 2018).

Who Is Italian Today?

> Mahmood, born and raised in Italy with an Egyptian father, isn't Italian but Lady Gaga, born and raised in New York with a Sicilian GRANDFATHER, is a source of Italian pride. I love my country. (neriitaliani.blackitalians Instagram post, 2019)

Ideas of who can be Italian have simultaneously appeared to expand and shrink in recent years. As observed by scholars such as Frisina and Hawthorne (2018), an expansion of multiple ways of being Italian is evident in the country's lived cosmopolitan realities (Werbner 2006). Dominant media, retrograde *jus sanguinis* citizenship laws and Italy's increasingly securitised governance, conversely, continue to shrink the possibility of who can be Italian, officially at least.

That Lady Gaga is a source of national pride while Mahmood's identity is regarded with suspicion is an indictment Italy's citizenship laws that poorly accommodate its residents. Under the present legislation, citizenship is still extended to the descendants of Italian emigrants who have 'Italian blood'—though may have never visited Italy—while it is denied to the Italian-born children of migrants until they turn eighteen, when they have a year to request citizenship, which is in any case conditional on uninterrupted residence in Italy. So when Alessandro Mahmood's song 'Soldi', won the 2019 Sanremo contest, the Italian media questioned whether he could really be regarded as Italian due to his father being Egyptian (Cazzullo 2019). The inclusion of Arabic verses in Mahmood's performance, moreover, angered far-right politicians, leading them to call for quotas of Italian music on the radio. It should be noted that calls for such quotas have never been made in response to the prevalence of English language music in Italy; pop songs with English lyrics have been played without controversy on Italian radio for decades.

In a country of cultural multiplicity, produced by Mediterranean crossings of people of all colours, creeds and ethnicities (Chambers 2008), why has this fantasy of ethnic homogeneity endured? And, how has it expanded so dangerously in a few short years? And, what is the relationship between Italy's populist regression to the far right and the so-called 'migrant crisis' of 2015? As Pugliese and Messina (2017) remind us, many of the peninsula's cultures, ethnicities and languages were disavowed and marginalised in the process of constructing Italy following its unification in 1861. Such disavowal, they argue, is 'foundationally constitutive of the heterogeneous and fractured assemblage that comes under the otherwise homogenising imprimatur of "Italian identity": namely, race and Whiteness.'[3]

Official Discourses of Italianness in a Fertility Campaign

Official discourses have contributed to the reification of a White homogenized Italian identity in recent years. In September 2016, for example, to mitigate Italy's declining birth rate, the Italian Minister of Health Beatrice Lorenzin launched a campaign to promote "good habits" for fertility (see Fig. 10.1). Advocating having children as "the best way to be creative," the campaign mobilized a series of poorly designed posters that encouraged Italians to start families young (Lorenzin, Fertility Campaign 2016). One poster promoted "correct lifestyles" as ones in which Whites have fun together (see Fig. 10.1). Invoking colonial era anti-miscegenation and white supremacist 'great replacement' ideology, an image of White people is juxtaposed with Black 'bad companions' it warns should be "abandoned" (see Fig. 10.1). Never mind that the image of the White folk featured was appropriated from a stock photograph previously used by a British dental campaign, featuring people with Anglo-Celtic ancestry remote from Italy.

As Italian Somali author Igiaba Scego pointed out at the time, Fertility Day was tied up in discourses of race that can be traced to the country's 1938 racial laws. These laws upheld the 'white race' as superior, a bizarre decree in any context, but particularly in Italy, a Mediterranean country

[3] According to Pugliese and Messina: "*South-verting* Italian Studies literally means freeing them from the North-normative pressure on which the very concept of Italianness is predicated. In this sense, our positionalities as émigrés, expats, exiles and first-, second- or third-generation diasporic subjects allow for the articulation of what is largely unspeakable within the territorial boundaries of the Italian Republic, reiteratively, the violent operativity of race in tracing the geo-biopolitical demarcation between the North and the South" (2017).

Fig. 10.1 'Promoting good habits, abandoning bad "companions". Correct lifestyles for the prevention of sterility and infertility.' (Source: Italian government minister for health campaign, Viral Facebook post, first accessed 22 September 2016)

that is inherently métisse, its regional cultures composed of people of diverse origins and phenotypes through the centuries (Giuliani and Lombardi-Diop 2013; Guglielmo and Salerno 2003). A slap in the face to Italian-born non-Italian citizen children (known as G2 for 'second generation'), the racialised overtones of Fertility Day both distracted and inadvertently drew attention towards citizenship rights for all those born on Italian soil, as the G2 poster depicts (Fig. 10.2).

Fig. 10.2 "Do you want more children? Acknowledge those you already have. Citizenship doesn't have an age. We've been waiting for 24 years." #reformcitizenshipnow G2 *Jus soli* citizenship campaign. (Source: Viral Facebook post, accessed 2 September 2016)

The tone-deaf campaign was widely attacked for recommending people to have children in the absence of jobs (with unemployment rates of up to 42%) and social support structures. The government's apparent lack of regard for the precarious labour market in which deep structural inequalities thwart the possibility of *una sistemazione* [establishing oneself], moreover, provoked outrage (Hu 2016; Iaccino 2016). Young people in Italy thus rallied to oppose the risible, insulting images depicting the 'right (white) way' to go about having children for the nation.

Sharing Outrage

A further example of normative Whiteness in Italy underpins the performance, especially online, of much racist anger towards Black Italians and migrants. This can be vividly observed in a series of manipulative distortions enacted in August 2017 by a White satirist, Luca Bottura, who retweeted an image of two famous Black North Americans, actor Samuel L. Jackson and basketball player Earvin Magic Johnson, who were holidaying in Italy. Johnson had tweeted a picture of the two Americans relaxing on a park bench in Forte dei Marmi, Tuscany, with several shopping bags from luxury outlets at their feet (see Fig. 10.3).

Claiming that he was undertaking a social experiment, Bottura reposted the image with the addition of stereotypes of African migrants inherent in the racist caption: 'Shame!', alongside the hashtags: 'resources, boldrini, salvini'[4] (see Fig. 10.4).

Millionaires Jackson and Johnson were thus readily coded as migrants with little to contribute, and a lot to take from the Italian economy. This image was soon appropriated in a meme which falsely declared the couple to be immigrants who had taken advantage of 35 euros given to them by the Italian state, and encouraged its affected audience to share if they too were outraged. As has been pointed out by Huddy et al. (2007), anger is a unifying force in groups and can be deployed to create collective identities. In this case, outraged viewers were called upon to identify as 'Italian victims' of a 'migrant invasion' (Fig. 10.5).

Bottura's 'experiment' relied upon hackneyed racialised assumptions of what it is to be Italian, encoded through a visual archive of European citizenship (Hall 1980; Meo 2018). It also tapped into the moral panic that migrants were squandering Italian taxpayers' resources, not unlike "that cliché of the South as a parasiting burden that voraciously consumes most of the resources produced by the North" (Pugliese and Messina 2017).

[4] Laura Boldrini was the President of the Italian Chamber of Deputies and was criticised for advocating openness towards refugees in Italy, while Matteo Salvini leads the xenophobic Italian Northern League party that promotes hatred towards migrants.

Fig. 10.3 Earvin Magic Johnson's tweet of himself and Samuel L. Jackson in Forte dei Marmi, Tuscany, 16 August, 2017

Fig. 10.4 'Shame! #resources #boldrini #salvini. Bottura's retweeted version of Johnson's Twitter post, 19 August, 2017

Fig. 10.5 Johnson's post was appropriated and retweeted as a meme that read: "Boldrini's resources in Forte dei Marmi. They shop at Prada with our 35 euro. Share if you are outraged!!!" Twitter, 19 August, 2017

The reality of the European Union granting Mediterranean arrival (first point of contact) countries such as Greece, Italy and Spain resources to administer these arrivals, so that the funds were not coming from Italian coffers, was poorly understood. In a post-truth era (Harsin 2015), media disinformation through the internet's rhizomatic channels contributed to the traction garnered by this item of fake news.

Widely shared on social networking sites and charged by their algorithms, the "truth market" (Harsin 2015) generated by the meme served to reproduce White Italian victimhood at the expense of African migrants. Few people recognised the men on the park bench as American celebrities on holiday and instead joined in the outrage at the vision of smiling African migrants 'living it up' at Italian taxpayers' expense (Oppenheim 2017). This fallacy was already been sustained elsewhere in Italian mainstream media outlets and on the social media accounts of far-right League leader Matteo Salvini who has claimed that "Africans don't belong in Italy"

[*'L'Africa in Italia non ci sta'*],[5] and that Italy is at risk of being colonised by migrants.

The status and wealth of these celebrities was thus rendered invisible by the racial codes internalized by the sharers of this tweet who demonstrated an inability to imagine Black people as anything other than poor migrants. This inability is in part a symptom of Italian media whitewashing their reporting of Italy's everyday lived realities—with African migration more often than not still aligned with criminality. It also indicates educative lacunas that continue to obscure the country's shared history with East Africa in the service of an essentialist national identity.

"Italians First, Work First"

Italians have been groomed for years by tabloid media to criminalise migrants and blame them for Italy's problems (Giustiniani 2003; Zinn 2011). League Leader Salvini's 2018 election campaign took advantage of common misunderstandings about immigration and its relationship to Italy's resources, as he deployed the populist rhetoric of putting Italians and 'work' first. Embedded in tropes of ethnic cleansing,[6] Salvini's campaign delivered a chillingly simple message: "get rid of *them* and I will look after *us*." Salvini promised to restore dignity to disenfranchised White Italians whose misunderstanding of migrants as a threat to Italy's labour markets led them to embrace his party's political brand of racialised divisions.

Salvini's hate speech has galvanised Italy's neo-fascist movements and sanctioned racialized acts of far-right violent extremism—including lone and organised attacks—that have targeted Black Italians and migrants (Mayr 2018). This was evident when far-right actors responsible for the spate of violent attacks against Black Italians in 2018 appeared to attract sympathy, rather than condemnation, from key League politicians who now govern the country.[7] Italy's borders are now more openly policed

[5] Salvini's assertion was made on the 9 January 2017 on the Mediaset television programme *Quinta colonna*, where it was met with applause. https://www.youtube.com/watch?v=XRYkobChyMg (Accessed 5 December 2018).

[6] One of Salvini's election promises was to "clean Italy's streets."

[7] The murder of Senegalese vendor Idy Diene on Ponte Vespucci in Florence in March 2018, the day after the Italian election, for example, was initially described in the media as a "misdirected suicide attempt." The murderer, Roberto Pirrone, claimed to be suicidal for economic reasons, but shot Diene multiple times instead of other White passersby or himself

than ever by far-right extremists whose multiplying graffitied emblems of fascism and neo-Nazism in public spaces are the tip of the iceberg of their threat to social cohesion and national security.[8]

Italian 'Good Sense'

Italy's former centre left Prime Minister Enrico Letta (2019) suggests that the current surge in attacks on migrants has its genesis in the unprecedented numbers of people arriving in Europe since 2015. Letta's 2019 manifesto for the regeneration of a progressive Italy, titled "*Ho imparato* [I've learned]", advocates for the need to speak sensibly about, and to accept, immigration in the country:

> One can only achieve [progressive social reform] with faces, ideas and arms of people who advocate for good sense, the Italian way. Yes, "Italian," because the persecution of immigrants isn't really very Italian at all. It has nothing to do with our culture and very little to do with our history. (Letta 2019)

To what brand of Italian good sense does Letta refer? And, what might it mean to refer to the persecution of migrants as un-Italian? The cultural assumption that Italians are good people, explored in historian Angelo Del Boca's *Italiani brava gente* (2005), arguably underpins Letta's invocation of Italian good sense. The last line, in which Letta asserts that the persecution of migrants has little to do with Italian history, reduces Italy's colonial past to the *molto poco* of a paragraph. How much indeed *has* Letta—taken here as representative of Italy's leftist ruling class—learned?

How can we make sense, then, of the predicament of Italy's histories of colonial violence and unacknowledged former colonial subjects and their children living in Italy, often as non-citizens? Within Italy's exclusionary citizenship regime, to what extent do the citizenship claims of those, for example, with heritage from Italy's former colonies such as Eritrea and

so as to be given a gaol [jail] sentence (Gostoli 2018). Pirrone was pitied in the press for having become depressed, frustrated and helpless at his lack of employment prospects. Even though Diene was shot six times, nowhere was the violence reported as an act of terrorism, as would have likely occurred had the murderer been a migrant (Gaye 2018).

[8]For a hopeful localised counternarrative to this far right reprisal, see the work of food artist Cibo: https://video.vice.com/en_us/topic/cibo

Somalia, complicate this picture and encourage redefinitions of national belonging?

The next section will explore these questions in relation to the 'Reimagining Citizenship' encounter held at the European University Institute in 2015, focusing on the interventions of a writer and filmmaker. It will then draw on an interview with a former Democratic Party MP to consider the citizenship law reform proposals that were approved by the Italian parliament's lower house in 2015, then obstructed in the Senate by thousands of amendments introduced by far-right parties until they were ultimately shelved in 2017 (Tintori 2018).

Reimagining Citizenship: An Encounter Between Researchers, Writers and Politicians

We must build a new kind of citizenship. (Kaha Mohamed Aden 2015)

Held in April 2015 at the European University Institute, the 'Reimagining Citizenship' encounter focused on investigating citizenship and national identity in Italy today.[9] Thinking with the work of Appadurai (following Anderson 1983), who has theorised "the image, the imagined, the imaginary" as key concepts that are inherently linked to communities, global developments and politics (1990, 2013), the encounter was premised on his crucial insight that imagination is a "social practice" (1990).

Paying special attention towards the predicament of migrants in Italy from the country's ex-colonies, the encounter provided a platform for Italian intellectuals of Somali background Cadigia Hassan, Antar Marincola and Kaha Mohamed Aden to discuss their writings on what it means to be an Italian with Somali heritage. The presence in particular of a politician from the Democratic Party who was working on reforming citizenship in Italy was designed to provoke a rethinking of citizenship in Italy that might provide policy directions. Following a screening of filmmaker Fred Kuwornu's *18 Ius Soli*, Italian MP Marilena Fabbri, Antonio Morone and Professor Anna Triandafyllidou discussed proposed Italian citizenship reform.

[9] Triandafyllidou, A. and V. Gerrand. Migration Working Group Special Session: 'Re-imagining citizenship: An encounter between researchers, writers and politicians.' https://www.eui.eu/events/detail?eventid=111413

There is now a significant body of scholarship exploring the ways in which writers with heritage from Italy's former colonies challenge ideas of a homogenous Italian national identity. Having written elsewhere on the ways in which representations of belonging and identity by Italo-Somali authors in particular reconfigure understandings of Italian history (see Gerrand 2008, 2016a), in this section I focus briefly on the interventions of the writers and the filmmaker. I then draw more detail on an in-depth interview with MP Marilena Fabbri to critically explore the proposed reforms to Italian citizenship laws. I consider these in light of Morone's insights to understand the relationship of the current *jus sanguinis* laws to the partitions enacted during the Italian colonial period.

Kaha Mohamed Aden's In-Between Understandings

Kaha Mohamed Aden's memoir, *Fra-intendimenti* (In-between understandings), was described at a seminar held in Naples as an "act of citizenship" (Aden 2010; 2011). Aden's work highlights the daily labour that still needs to be performed by women migrants of colour to belong in Italy as Italian. For example, Aden's book explores some of the ways in which she is still often regarded as a foreigner in her hometown of Pavia, in spite of her long-term residence there and fluency in the Italian language (which she even speaks with a regional accent). As a Somali woman in northern Italy, ordinary belonging can be an ordeal, requiring ongoing exhausting efforts. When she is not mistaken for a prostitute, Aden's identity is frequently subjected to interrogation or policing. Aden has also been a target for well-meaning feminist advice (Carroli and Gerrand 2012; Gerrand 2016a).

At the 'Reimagining Citizenship' encounter, Aden begun her talk by referencing the recent drowning in the Mediterranean of unprecedented numbers of people seeking to reach Europe. This was the beginning of what would soon be problematically referred to as the "migrant crisis" (De Genova 2018). Aden could not think of any subject more critical for discussion than the question of innocent people drowning at sea and the cruel indifference of the systems of fortified borders put in place that have led to people with the some ("wrong") passports (Scego 2016, 2019) to travel unsafely by boat rather than taking, for example, a low-cost flight to

Jus soli – birthright citizenship

J. Sanguinis – nationality decided by nationality of one or both parents.

J culturae – to stay in country (>2y) and complete educational courses etc.

Europe. Towards the end of the encounter, I asked Aden to discuss her book in relation to reimagining citizenship in Italy.[10]

> VG: How did you come to choose this title? What does it represent?
>
> KA: I wanted to underscore the question of responsibility because people can understand or not understand each other. In my stories there are characters that in a complicated way cannot understand one another, not solely due to a lack of ability or resources. Usually ignorance is the explanation given. I suggest that there's something else that we might call the willingness to move towards the other. Actually, moving towards the other means finding a space of encounter that is that hyphen that sits in the middle: what you English speakers refer to as 'in-between', no?
>
> The 'between' is this place, a place where people meet, listen and construct a new citizenship. This is the way it is because, if Somalia has taught me anything, the issue of citizenship doesn't exist in any definitive sense, not even citizenship by blood. Because I was part of a community when the Somali state existed and there was a dictatorship too. When this dictatorship was overthrown, there was a return to clannism that is linked to the patriarchal lineage and thus to blood. At one time I found myself not part of that thing connected to blood. And then, when the war broke out, I was in Italy. My problem then became one of not being able to access Italian citizenship, as Italian citizenship is linked to blood ties, *jus sanguinis*, which is the reason we're speaking here today.
>
> My thesis was and remains that we must, through a big effort of will and the work of reciprocal listening, build a new kind of citizenship. But first we must understand that in reality it is missing and that we really desire it. There's a story in which I recount that I would like to look out on a balcony, start building a home with a balcony suspended above the Indian Ocean, built with stories and reciprocal listening, and to hear the tales and the iodine. We must work in this direction and it's possible because humanity has done it thousands of times before. It has also failed thousands of times but it can be done, why not? (Interview with Gerrand and Sabrie 2015)

Aden highlights the critical role of listening and interaction in breaking down racialised structures and assumptions that underpin Italian citizenship laws. Gravitating towards a space that is neither here nor there, but perhaps here *and* there, enables a reconfiguring of identity and a shift towards an appreciation for shared complexities. The in-between space, signified by the hyphen, long examined in research on identity, is a critical

[10] The interview was filmed by Asha Sabrie.

part of how humans find ways to coexist in their diversity. Kaha's intervention in this encounter affirms the importance of this understanding and its ongoing relevance to understanding how to imagine the futures we need for a convivial existence that extends beyond racialisation, colonisation and nativism.

Cadigia Hassan's 'La Bambina Salvata dal Coccodrillo'

Writer, journalist, and founder of the Padua-based Migrant Women's Network, Cadigia Hassan discussed her experiences of growing up Black in Italy. She sent me a copy of her short story, "La Bambina Salvata dal Coccodrillo," which sets a Somali-inspired folk tale that takes place across the contexts of Somalia and the Euganean hills of the Veneto region (Hassan 2013). The crocodile in this tale becomes a friend to the child protagonist, and a spirit guide that connects her to her Somali heritage. Hassan's upbeat attitude towards her multiple affiliations, and the lack of contradiction between them, appeared in contrast to some of the other speakers who emphasised the weight of their struggles.

Hassan's illustration of her identity, lived dynamically both in Italy and in Somalia, reveals the everyday lived upbringing and continuity of Somali and Italian cultures in this little girl's Italian life. We might view the title of the story, which expresses the notion of the crocodile saving the child, as a metaphor for the little girl's Somali heritage that redeems her by allowing her to feel pride about her mixed-race identity.

Fred Kuwornu's *18 jus soli*

Through interviews with a series of Italian-born young adults, Kuwornu's compelling documentary highlights the dramatic limitations imposed on the lives of young people living in Italy with origins outside the country.[11] The injustice is clear; the solution, equally so. What is preventing these citizenship law reforms? These young people, from a wide range of circumstances, speak in regionally inflected Italian accents; they embody their birthplace. From the way they engage, and by using close-up montage, the documentary highlights their everyday, ordinary, belonging to

[11] For an incisive analysis of Kuwornu's documentary, see the concluding chapter of Teresa Fiore's 2017 monograph *Pre-Occupied Spaces: Remapping Italy's Transnational Migrations and Colonial Legacies.*

Italy's regional communities. This ordinariness serves to throw into doubt current citizenship laws and underscore how and why they are not just outmoded, but punishing. These young people, the documentary emphasizes, have as much right to a future on Italian soil as other Italians (Kuwornu et al. 2012; Leal Riesco 2013).

Kuwornu highlights the history of Giorgio Marincola in the documentary. The question raised by Kuwornu in the documentary, about how Italy can ignore its partisan history in which the role of Italian Somali partisan Giorgio Marincola, was central. Given that this is also the subject of Antar Marincola/Wu Ming's book, *Timira*, Antar Mohamed's intervention at the 'Reimagining Citizenship' encounter critically connected the documentary about citizenship with his family's legacy. In particular, Antar's discussion of his book elucidated Italy's constitutive colonial entanglement with Somalia.

Antar Mohamed Marincola's *Timira*

Timira: Romanzo Meticco (*A Métisse Novel*) tells the story of the Marincola family through the eyes of Black Italian protagonist Isabella Marincola (or Timira, in Somalia), whose son co-wrote the book with Wu Ming (Wu Ming and Mohamed Marincola 2012). The child of an Italian officer and a Somali mother, like many Italian Somali so-called métisse children growing up in colonial Somalia (Morone 2018), Isabella was sent to live in Italy with her brother, Giorgio.

It was common to send Italian Somali children to live in Italy, especially after the end of the 10-year Italian trust administration of Somalis (AFIS Afis) in 1960, when Somalia became independent. As Morone points out, at this this time, Italian Somalis were excluded from wider Somali society, having been educated in Italian-run Christian institutions. Somali *jus sanguinis* citizenship laws introduced in 1963 stipulated that to be Somali, one had to be the child of a Somali father. This led to the potential statelessness of this generation of children, who had little choice but to migrate to Italy and apply for Italian citizenship (Morone 2018).

Once in Italy, rather than being accepted as Black Italians, Italian Somalis once again faced marginalisation, this time in the form of discrimination and racism based on the colour of their skin. In spite of their ready acquisition of Italian citizenship, many were treated as second-class citizens (Morone 2018). In addition, upon leaving Somalia, many of these children lost their relationship with their Somali mothers, who stayed behind as their children were forcibly removed, as was the case for Isabella Marincola.

Timira includes the history of Italian Somali partisan Giorgio Marincola, who fought in the Italian Resistance movement against Fascism. A central figure in Italian history, who has his origins in Italy and Somalia, Marincola exposes Italy's entanglements with Somalia, opening a critical window into its hitherto ignored and silenced histories, thereby restoring the centrality of the Horn of Africa to Italian identity.

Historian Antonio Morone's intervention in the workshop framed the reluctance to acknowledge Italian-born children as in keeping with Italian racialised colonial genealogies. Morone's scholarship addressed the complex history of Somali *meticci* and their journey from Somalia to Italy (2015, 2018). During the 'Reimagining Citizenship' encounter, Morone discussed the multisided colonial legacy and the racist roots of contemporary Italian citizenship.

What is clear from this encounter is the extent to which citizenship law in Italy has neglected to fully engage with the complexities and lived realities of Italy's colonial legacy. In the next section, an in-depth interview with Marilena Fabbri will shed light on a government perspective of citizenship law reform. While very well-intentioned, this government perspective ignores the particular genealogy of East African descendants in Italy and their potential to galvanise an official reimagining of Italian identity in order to successfully reform Italian institutions of citizenship.

Former Member of Parliament Marilena Fabbri on Italian Citizenship Law Reform: A Government Perspective

> Italy became a cradle of culture because ... it was a point of encounter between peoples that came together sometimes violently and contaminated one another. The notion of an 'Italic race' is completely ahistorical, anachronistic and unfounded. There's no place in Italy that could claim [racial] purity of any kind. This is Italy's strength. (Interview with Marilena Fabbri 2019)

The proposed law was indeed a compromise—no unconditional *jus soli* was on the table. Fabbri intervened in our workshop to speak about these proposed reforms. Three years later, I interviewed her on why the reforms failed. A mediation between various interests at play, the proposal to reform Italian citizenship law from *jus sanguinis* to *jus soli* attempted to strike a balance between those who aspired to simplify the laws in order to recognise the right to citizenship on the one hand and those who questioned and feared the recognition of this right on the other hand. The centre left coalition sought not *jus soli*, but rather *jus soli moderato* and *jus culturae* as a compromise. Fabbri still regards it as an excellent compromise.[12]

Fabbri, who now works for the Emilia Romagna region where we met, spoke of the incredible opportunity she was given to work on these citizenship law reforms, reforms in which she "really believed" and to which she was profoundly committed. Fabbri articulated some of the very sophisticated measures at which the proposals arrived, the particular details in which they would be enacted that would respect the rights of prospective Italian citizens, and the fears of some Italian citizens.

> The work we did was an incredibly important political battle and we undertook it with utmost attention to detail. I read all the citizenship law proposals that were lodged. There were 25 proposals lodged, including one from a popular initiative. Almost all of them, except for one,[13] were proposals that opened up the legislation. They proposed to improve the 192 law by simplifying access to citizenship acquisition. (Interview with Fabbri 2019)

Some of the key considerations that were included in the proposed reforms were extending the time frame (from one to two years) for which those turning eighteen to request citizenship, altering the requirements for proving continuity of residence in Italy, and enabling children who arrived in Italy before they turned eleven to acquire Italian citizenship once they completed a cycle (five years) of schooling (*jus soli temperato*).[14] The proposals were designed to amend the existing 192 legislation, number 90 of 1992, for adults and minors. The compromise that was struck was intended to amend the law solely for minors, the children of first-generation immigrants.

This was the first compromise because the centre right parties did not want the government to amend the law for adults. For the right it was to

[12] Interview with Marilena Fabbri, 18 February 2019.

[13] Interview with Fabbri 2019.

[14] Interview with Fabbri 2019.

remain a matter of 'naturalisation' of foreigners who had arrived in Italy, especially those from outside Europe who must wait at least ten years before being eligible for Italian citizenship. Many, however, have been living in Italy for well over ten years, but are not eligible due to interruptions in their continuity of residence—interruptions such as loss of work reset the counter.

In reality though, the proposals were not just concerned with Italian-born children. They also were to be applied to anyone who had arrived in Italy before their tenth birthday and who had continued to reside in the country. It therefore was not limited to minors, but also included adults who had arrived in Italy before they turned ten and grown up in the country. There were approximately one million potential minors who could have benefitted from the law—not all at once, as some members of the right falsely contended, but once they had fulfilled the particular criteria.

The *Jus Culturae* Provision

I asked Fabbri about the *jus culturae* provision of the law. This provision addressed all children who had been born in Italy, even if their parents did not hold a long-stay foreigners' permit but who may have had a work, study or humanitarian permit. In these cases, the children of regular foreigners or those who had arrived in Italy before turning ten and who had attended Italian schools for at least five years were eligible. No graded school certificate was necessary other than a primary school certificate of attendance attesting to the positive completion of the study cycle at the end of the fifth grade, when students are sent to secondary school.[15]

After five years of school attendance, the government would only consider whether the cycle of schooling had been completed. Fabbri emphasized her party's decision not to link the possibility of citizenship acquisition to a particular student's grades. This became the subject of heated debate between the centre left, who agreed with this position, and centre right, who believed the grades should be taken into consideration. Fabbri and

[15] It was also sufficient to have completed two years of primary school, for example, and three years of secondary school. Provided the schooling began before the child turned 11, it would have been possible to request citizenship.

her colleagues sought to avoid a situation in which Italy would only accept the best of its students as citizens, which would have put pressure on teachers and schools to grade students without citizenship differently, especially if students did not perform adequately but teachers did not wish to negate them their right to Italian citizenship. Fabbri acknowledged such a situation would have created problems and dynamics that were inappropriate in an education system. Thus school attendance, rather than performance, was the proposed criterion for citizenship acquisition (Interview with Fabbri 2019). School attendance, Fabbri affirmed, is what leads to "cultural acquisition and integration."

The requirement of school attendance was to be extended to adults who had arrived in Italy before the age of ten and who had not had problems in the Italian justice system or posed a security risk to the state. These adults would have been invited to apply to become Italian citizens. The discourse of education producing social inclusion runs through Fabbri's discussion of the proposed reforms. Integration was a key component behind why ten was chosen as an age limit. (Whoever arrives in Italy before ten must go to school until they are at least sixteen years old.) In this time, a child becomes part of at least five years of school-based integration.

Jus Soli Temperato Citizenship as an Investment

> Education is an onerous public investment. Italy invests funds in educating these young people. So it's completely myopic and wasteful, not even taking into account other elements of value, to think of not encouraging these kids who have been formed by the Italian system to remain in Italy. These young people have been formed and trained to work in Italy and have the resources to contribute to the development of our country. (Interview with Fabbri 2019)

Economic investment was thus another key component of the Democratic Party's considerations in their proposed citizenship amendments. The Italian state is already investing significantly in the education of these children. It would have been short-sighted on the part of the Italian state to invest in shaping these children's futures, and then to force them to leave the country to pursue work abroad. Whether they have foreign parents or not, Fabbri emphasizes, the reality is that these students are growing up and being educated by the Italian state. For this reason, the reforms included recognition of adults who had been educated entirely within the Italian system, some of them with university degrees.

Demographic Considerations

> Through Istat,[16] we verified that 140,000 young people had the characteristics of having been born, or having grown up, in Italy. This cohort could take advantage of the *jus culturae* criteria. 95 per cent of them were between 20 and 45 years old, the most productive time in life both in terms of work and reproduction. This was also a highly educated cohort, many with university qualifications, challenging the misconception that migrants are poorly educated. (Interview with Fabbri 2019)

Another critical consideration was Italy's declining birth rate. The Democratic Party saw these prospective Italian citizens as critical to its demographic stability and growth. There was a clear rationale to keep highly educated adults between 20 and 45 years of age in the country for their capacity to grow Italy's economy and to contribute to its demography.

Public Support for Citizenship Law Reform

> I believe it was theft ... a minority was robbed [of their right to citizenship] by a majority of citizens. And this denial happened on the basis of fear which produced and instrumentalised the politics of closure. This involved the idea that giving rights to some removes rights from others. (Interview with Fabbri 2019)

I asked Fabbri about 2012, when Giorgio Napolitano claimed it was folly not to consider the children of immigrants as Italian citizens. At that time, Istat had registered a public consensus that it was time to reform Italy's citizenship legislation. Fabbri spoke about how the law began as a people's initiative backed by 250,000 Italian citizens. The signatories came from all over Italy and included a group of associations such as 'I'm Italy too', and some that were tied to the Catholic Church, an active supporter of the reforms (Tintori 2018; Zinn 2011). Such reforms would not bring about religious tensions in Italy, as the far right had argued.

According to Fabbri, the far right has misunderstood identity as a zero-sum game, emphasizing the notion of identity loss. Even though Italy's

[16] Italian National Institute of Statistics 2017.

Muslim population comprises approximately 1.5 per cent of the population, this idea of identity loss was underscored particularly in discussions of Islam, which the far right argued constitutes a colonisation of the country due to ubiquitous and self-serving misunderstandings of the religion. (Interview with Fabbri 2019) Fabbri was puzzled by the fear of a country of almost 60 million people being overtaken or 'contaminated' by 1.5 per cent of its population.

Increased Competition?

The far right furthermore capitalised on the notion of increased competition as a threat to Italy's economic resources. In a time of economic crisis, the right was able to use the idea that making citizenship acquisition easier would disadvantage Italian citizens by increasing the number of people using Italian services and taking Italian jobs. This ignored the fact that whether or not foreign residents are citizens of Italy, they are nonetheless legally entitled to work and make use of Italian services such as education, hospitals and housing (Zinn 2011).

As Fabbri confirmed, "Economic rights are acknowledged before citizenship. Access to schooling, health and public housing and work is separate from citizenship" (Interview with Fabbri 2019). Even though increased acquisition of Italian citizenship would make little difference to the allocation of resources in the country, the far right nonetheless successfully distorted this issue for political gain at a time of economic downturn.

Jus Soli: Towards Social Inclusion

> What would the amendments to law 192 have achieved? They would have improved integration and the sense of belonging of these people. (Interview with Fabbri 2019)

Fabbri regards it as important to ensure children born in Italy can acquire citizenship before they enter secondary school:

> It may not matter whether a child has Italian citizenship in infancy and in the early years of schooling, but when the identity of that child begins to form as they enter adolescence it becomes crucial. That is the moment when

a child begins to ask questions and experience difference and discrimination. (Interview with Fabbri 2019)

Fabbri believes that Italy takes a risk when failing to invest in the people who will build the country's future:

> These young people live and breathe multiculturalism. They're already cultural and commercial mediators, bridges connecting their countries of origin because they're sources of hybrid culture—that of their parents and the culture in which they have grown up. So they're definitely a bridge. (Interview with Fabbri 2019)

Instead of appreciating this generation of young adults' potential and the cultural richness they bring, Fabbri laments the fact that this generation continues to experience social exclusion. This is not just short-sighted, but unjust. Fabbri affirms that Italy is what it is thanks to its ethnocultural 'contaminations': Italy's multi-faceted wealth of architectural, culinary, landscape and fashion innovations, in her view, should be valorised and built upon, as they have been in the past, to make Italy competitive again, in spite of its small presence on the global scale.

Interestingly, Fabbri's Mediterranean "contamination" discourse did not explicitly extend to the legacy of Italy's colonial expansion in the Horn of Africa, suggesting that even among the most progressive forces in government, it is a commonplace to acknowledge Italy's genesis in 'contamination', but its relatively recent history of colonial expansion rarely figures into descriptions of this contamination.

It should be noted that no special distinction was afforded to migrants from Italy's ex-colonies in the proposed *jus soli temperato* citizenship reforms. In spite of Italy's long shared history with countries in the Horn of Africa such as Ethiopia, Eritrea and Somalia, where Italian can still be witnessed in a variety of customs, cultural and linguistic practices, no differentiation between migrant groups has occurred in understanding the prospective 'integration' capacity of these migrants and their children. This lack of differentiation might be viewed as another symptom of the failure to officially recognise the legacy of Italy's colonial past, a lacuna that continues to shape the country's political and educational institutions.

Has this failure contributed to the coalition of far-right forces' success in parliament? While the discourses need not be mutually exclusive, the rhetoric of contamination may obscure more than it enlivens for understanding the specific dynamics associated with the nation's enduring colonial entanglements. The elision of Italy's colonial history that occurs in this deployment draws attention once more to the importance of reclaiming the Black Mediterranean.

Nowadays there is more likely to be a consensus against reforming Italian citizenship laws among Italian citizens. The current system continues to produce irregularity, punishing people rather than assisting them in finding stability. Not unlike the treatment of the so-called *meticci* in Somalia, who were segregated from the customs of their country of birth while being at the mercy of Italian missionary institutions that led many into poverty and prostitution (Morone 2018), Italy's citizenship and immigration laws tend to subjugate rather than give migrants a platform for upward mobility by granting them all manner of rights.

Remedying the Current Catch 22

Under the current laws, implemented under the 2002 Bossi Fini amendments which produce irregularity, a catch-22 situation is not uncommon among migrant families: without a regular work contract, a family will not be eligible for a foreigner permit. Without a foreigner permit, it is almost impossible to find a regular work contract. This state of affairs seems to be designed to punish migrants and their children:

> The economic crisis puts migrant families in a double bind because it throws them into an unregulated state. What happens to the children in these families? If the family loses its stay permit at the same time that it loses its work contract, it will lose its legal right to reside in Italy because without a stay permit you cannot sign up at the Registry Office. This means the legal continuity of residence is interrupted. So even if the child of migrants knows he has always lived in Italy, has attended school and is 18, if he or she is unlucky enough to be part of a family that in the meantime found itself caught up in this situation, the child would not be able demonstrate their continuity of residence. When this happens, the child is treated like any other migrant who may have arrived on Italian soil six months ago. (Interview with Fabbri 2019)

The former integration minister Cécile Kyenge led the Democratic Party to resolve this residence problem by enabling migrants to demonstrate their continuity of residence in a variety of ways—such as through attending school, university or gaining a diploma—instead of needing to rely upon legal residence. This meant that children's proof of residence was no longer tied to their parents' work situation.

> We were going to widen the window of time so that at 18, young Italian-born children of migrants could request citizenship within two years. We offered a slightly longer margin of time in which to decide. We also proposed to amend the rules around continuity of residence. To be eligible for citizenship at 18, a person must have lived in an uninterrupted way on Italian soil since birth. There are many families who have been living in Italy for twenty or thirty years who have been affected by the economic downturn and lost their jobs without being able to find other work within two years of losing their jobs. If a migrant loses his or her job and their stay permit simultaneously expires, they have a year in which to find a new job. If, within that year, they don't find a job, it's not possible to request another stay permit. Without a stay permit, however, it's impossible to find a job. Without a job no stay permit. This is an absurd requirement that produces situations of irregularity and does not permit someone who had regular job but lost it to return to regular work. (Interview with Fabbri 2019)

Thus a new flexibility was to be extended to adults in such precarious situations, who would have the opportunity to deploy various types of evidence that they had had continuity of residence, including previous periods of legal residence in Italy in catch-22 circumstances. Fabbri gave the example of a family that may have arrived in Italy regularly, but stayed with friends or their employer as they did not have a home. These workers might have found themselves in this predicament for some years: years that would not have counted towards their residency in Italy. Once the family improved their situation and found their own accommodation, they would be able to obtain a residence permit and would be considered to be legally residing in Italy.

In the current system legal, residence does not recover previous years of living in Italy: it is only formalised from the time in which a family joins the Registry Office. If a family moves to another town and does not sign up to the registry of that town, it will once again lose its continuity of residence as it will not be linked to the previous residence and will therefore record the family as newly arrived. The proposed amendments would have counted residency from the first issue of the stay permit, and would have automatically recovered previous stay permits as well as allowing other forms of proof to be valid, such as a health card or proof of school enrolment.

Fabbri believes it is unlikely that the present political situation will allow this discussion to be reopened, especially not before the European elections in May 2019.

> If Europe confronts the immigration issue seriously and correctly, I believe that will reopen the possibility of discussing citizenship reasonably and rationally again. It would be even better if there were a discussion at the European level. When she was a minister, Cécile Kyenge was trying to open a discussion on European citizenship—to identify a set of elements that were common at the European level because the reality is that citizenship from one European nation is the doorway to a European passport. This can be a delicate subject because citizenship acquisition is a sensitive topic as we've seen. (Interview with Fabbri 2019)

Conclusion

This chapter is based on an encounter that was held at the European University Institute in April 2015 which sought to re-imagine citizenship in Italy through dialogue between Black Italian intellectuals, researchers and a politician. To better understand the relationship between citizenship and national identity, I analysed a series of recent images that characterise the exclusivist and often racialized perspectives on who can be Italian today. I argued that such perspectives provide a foundation for the rising prominence assumed by white supremacist neo-fascism and far-right violent extremism in the country.

The second part of the chapter reviewed the reimagining citizenship encounter in which participants discussed how Italian citizenship laws have circumscribed the lives of Black Italians and migrants in Italy and its colonies. An interview with a former Democratic Party MP who was instrumental in designing the citizenship reforms and who was present at the encounter elucidates the proposals and obstacles to reforming Italian citizenship law from *jus sanguinis* to *jus soli*.

A lack of consideration given within these proposals to the children of former colonial subjects suggests the extent to which there is still no common appreciation of Italy's shared past with Eritrea and Somalia in particular. Africans and Muslims are thus still viewed as 'Others' in Italy, rather than in a longstanding relationship and interaction as cultural, linguistic and proximal neighbours (Gerrand 2008).

Future proposed changes to the 1992 law might account for these shared histories. They could build on some conditions that acknowledge the cultural commons shared by Italy and the East African countries it colonised and therefore acknowledge that what it means to be Italian today also has its roots in the Horn of Africa. Those from the Horn are not from 'outside'; they are longstanding cultural insiders. Wide acknowledgement of their 'insider' status should be included in future Italian citizenship reforms not solely from the fact of time spent on Italian soil and in Italian schools in Italy, but also time spent in Italian institutions, for example those in Somalia that produced a generation of speakers of perfect standard Italian. The people who spent time on Italian soil or in Italian schools and institutions play a role in Italian history that complicates essentialist reductions of Italian identity to particular bloodlines or ethnicities. The visibility conferred by such acknowledgement would do much more than privilege the recognition of Italy's former colonial subjects on Italian soil. By representing them as cultural protagonists, it would serve the affiliations of all Black Italians and migrants in Italy who are inherent, rather than extraneous, to Italian and Mediterranean identity.

References

Aden, K.M. 2010. *Fra-intendimenti*. Roma: Nottetempo.

Aden, K. M. 2011. Atti scritturali come atti di cittadinanza: 'autoritratto'. Università degli studi di Napoli 'L'Orientale', 7 April.

Amin, A. 2013. Land of Strangers. *Identities* 20 (1): 1–8.

Anderson, B. 1983. *Imagined Communities: Reflections on the Origin and Spread of Nationalism*. London: New Left Books.

Appadurai, A. 1990. Disjuncture and Difference in the Global Cultural Economy. *Theory, Culture, Society* 7: 295–310.

———. 2013. *The Future as Cultural Fact: Essays on the Global Condition*. London: Verso.

Balibar, E. (Trans. J. Swenson). 2004. *We, the People of Europe? Reflections on Transnational Citizenship*. Princeton: Princeton University Press.

Bauböck, R. 1999. *Recombinant Citizenship*, Political Science Series No. 67. Vienna: Institute for Advanced Studies.

Bauman, Z. 2017. *Retrotopia*. New York: Wiley.

Camilli, A. 2018 'Cosa prevede il decreto sicurezza e immigrazione?' *Internazionale*, 27 November.

Carroli, P., and V. Gerrand. 2012. *La mia casa è dove sono*: Subjects and Narratives Beyond National Borders. *Scritture migranti: rivista di scambi interculturali* 5: 81–104.

Cazzullo, A. 2019. Mahmood e l'inevitabile integrazione che arricchisce l'Italia. *Il Corriere*, February 10.

Chambers, I. 2008. *Mediterranean Crossings: The Politics of Interrupted Modernity*. Durham: Duke University Press.

Colombo, E., L. Domaneschi, and C. Marchetti. 2011. Citizenship and Multiple Belonging. Representations of Inclusion, Identification and Participation Among Children of Immigrants in Italy. *Journal of Modern Italian Studies* 16 (3): 334–347.

De Genova, N. 2018. The "Migrant Crisis" as Racial Crisis: Do Black Lives Matter in Europe? *Ethnic and Racial Studies* 41 (10): 1765–1782.

Del Boca, A. 2005. *Italiani, brava gente?* Milan: Neri Pozza.

Finotelli, C., M.C. La Barbera, and G. Echeverría. 2018. Beyond Instrumental Citizenship: The Spanish and Italian Citizenship Regimes in Times of Crisis. *Journal of Ethnic and Migration Studies* 44 (14): 2320–2339.

Fiore, T. 2017. *Pre-Occupied Spaces: Remapping Italy's Transnational Migrations and Colonial Legacies*. New York: Fordham University Press.

Frisina, A., and C. Hawthorne. 2018. Italians with Veils and Afros: Gender, Beauty, and the Everyday Anti-Racism of the Daughters of Immigrants in Italy. *Journal of Ethnic and Migration Studies* 44 (5): 718–735.

Gaye, C.T. 2018. Il silenzio "negazionista" della politica. *Il Manifesto*, March 14.

Gerrand, V. 2008. Representing Somali Resettlement in Italy: The Writing of Ubax Cristina Ali Farah and Igiaba Scego. *Italian Studies in Southern Africa* 21 (1): 270–295.

———. 2016a. *Possible Spaces of Somali Belonging*. Carlton: Melbourne University Press.

———. 2016b. Migration and Resettlement Patterns of Somalis in the Diaspora: A Case Study of Italy and Australia. *Somali Studies: A Peer-Reviewed Academic Journal for Somali Studies* 1: 50–76.

Gillio, G.M. 2017. Soli, senza ius. *Articolo 21*, January 24. https://www.articolo21.org/2017/01/soli-senza-ius/

Giuliani, G., and C. Lombardi-Diop. 2013. *Bianco e nero: storia dell'identità razziale degli italiani*. Firenze: Le Monnier Università, Mondadori Education.

Giustiniani, C. 2003. *Fratellastri d'Italia: vite di stranieri tra noi*. Bari: Laterza.

Goodhart, D. 2017. *The Road to Somewhere: The Populist Revolt and the Future of Politics*. London: C. Hurst & Co Publishers.

Gostoli, Y. 2018. Protests and Questions Over Killing of Senegal Migrant Idy Diene. https://www.aljazeera.com/news/2018/03/protests-questions-killing-senegal-migrant-idy-diene-180308130543754.html

Gugliemo, J., and S. Salerno, eds. 2003. *Are Italians White? How Race Is Made in America*. New York/London: Routledge.

Hall, S. 1980. Encoding/Decoding. In *Culture, Media, Language: Working Papers in Cultural Studies, 1972–79*, ed. D. Hall, A. Lowe Hobson, and P. Willis. London: Hutchinson.

Harsin, J. 2015. Regimes of Posttruth, Postpolitics and Attention Economies. *Communication, Culture & Critique* 8: 327.

Hassan, C. 2013. *La bambina salvata dal coccodrillo*. Self-published, Padova.

Hu, C. 2016. Italy Begs Women to Have More Babies for the Sake of the Nation, Totally Embarrasses the Nation. *Quartz*, September 1.

Huddy, L., S. Feldman, and E. Cassese. 2007. On the Distinct Political Effects of Anger and Anxiety. In *The Affect Effect: Dynamics of Emotion in Political Thinking and Behavior*, ed. G.E. Marcus, W.R. Neuman, M. MacKuen, and A.N. Crigler. Chicago: University of Chicago Press.

Iaccino, L. 2016. Italy Fertility Day Backfires with 'White People Good, Black People Bad' Booklet. *International Business Times*, September 22.

ISTAT, 2017. Italian National Institute of Statistics Annual Report 2017, Rome, Published 17 June.

Khrebtan-Hörhager, J. 2019. Italians First: The New Borders of European Humanity. *Europe Now: A Journal of Research and Art*, February 5.

Kress, G. 2009. *Multimodality: A Social Semiotic Approach to Contemporary Communication*. London: Routledge.

Kuwornu, F.K., M. Giacomel, L. Iacovacci, G. Marincola, L. Levy, S. Ponzio, V. Ag, and C. Botsford. 2012. *18 Jus Soli*. http://www.worldcat.org/title/18-ius-soli-il-diritto-di-esseri-italiani/oclc/828204120. Produced in Collaboration with Struggle Filmworks and Associazione amici di Giana.

Leal Riesco, B. 2013. Review of Fred Kuwornu's *18 Ius Soli*: The Right to Be Italian. https://www.okayafrica.com/african-film-18-ius-soli-fred-kuwornu/

Letta, E. 2019. *Ho imparato*. Bologna: il Mulino.

Lorenzin, B. 2016. Iniziative volte a sostenere la natalità, con particolare riferimento agli interventi di carattere economico e sociale e alle misure per la prevenzione e la cura dell'infertilità. (Initiatives designed to promote births, with particular reference to economic and social interventions and measures to prevent and manage infertility.) Address by the Health Minister to the Chamber of Deputies, Sitting n. 3-02507. 28 September.

Mayr, W. 2018. The Dangerous New Face of Salvini's Italy. *Spiegel Online*, December 12. http://www.spiegel.de/international/europe/matteo-salvini-has-emboldened-fascists-in-italy-a-1243164.html. Accessed 16 Feb 2019.

Meo, E. 2018. I *codici estetici* della cittadinanza: *Visualità* e razzismo nel dispositivo europeo della cittadinanza. In *Visualità & (Anti)Razzismo*, ed. E.A.G. Arfini, V. Deplano, A. Frisina, G. Giuliani, M. Ghebremariam Tesfaù,

V. Perilli, A. Pes, T. Petrovich Njegosh, G. Proglio, and D. Salerno e A. Surian. Padova: University of Padova Press.

Morone, A.M. 2015. L'Italianità degli altri. Le migrazioni degli ex sudditi coloniali dall'Africa all'Italia. *Altreitalie* 50 (gennaio-giugno): 71–86.

———. 2018. Gli italo-somali e l'eredità del 'colonialismo. *Contemporanea* 2: 195–222.

Moual, K. 2012. Napolitano: "Cittadinanza ai figli di immigrati". *Il sole 24 ore*, December 19.

Nadotti, C. 2018. Crescono gli studenti non cittadini italiani. *La Repubblica*, marzo 29.

O'Halloran, K., and B.A. Smith. 2011. *Multimodal Studies: Exploring Issues and Domains*. Routledge, New York: Oxfordshire.

Oppenheim, M. 2017. Samuel L Jackson and Magic Johnson Mistaken for "Lazy Migrants" by Italians After Shopping in Tuscany. *The Independent*, August 21.

Papacharissi, Z. 2010. *A Networked Self: Identity, Community, and Culture on Social Network Sites*. London: Routledge.

Pesarini, A. 2017. "Blood Is Thicker Than Water": The Materialization of the Racial Body in Fascist East Africa. In I. Fattacciu and C. Fogu, eds. *Zapruder World: An International Journal for the History of Social Conflict* 4. https://doi.org/10.21431/Z33S32

Proglio, G. 2018. Is the Mediterranean a White Italian-European Sea? The Multiplication of Borders in the Production of Historical Subjectivity. *Interventions* 20 (3): 406–427.

Pugliese, J., and M. Messina. 2017. South-verting Italian Studies: An Introduction. *Muiraquitã—Revista de letras e humanidades* 5 (2): 1–10.

Queirolo Palmas, L. 2006. *Prove di seconde generazioni: giovani di origine immigrata fra scuola e spazi urbani*. Milano: FrancoAngeli.

Sardelic, J. 2019. *Racialised Citizenship Regimes: Minority Statelessness in the International Order of States*. March: International Studies Association Convention in Toronto.

Scego, I. 2016. Viaggiare è impossibile se non sei nato nel paese giusto. *Internazionale*, April 11.

———. 2017. La cittadinanza italiana ai figli di migranti è una conquista per tutti. *Cosmopolitan*, February 28.

———. 2018. Italy Is My Country—But It Must Face Its Racist History. *The Guardian*, September 16.

———. 2019. Aprite gli aeroporti. *Internazionale,* March 12.

Stevenson, N. 2003. Cultural Citizenship: Cosmopolitan Questions. Berkshire: Open University Press.

Tintori, G. 2018. Ius Soli the Italian Way. The Long and Winding Road to Reform the Citizenship Law. *Contemporary Italian Politics* 10 (4): 434–450. https://doi.org/10.1080/23248823.2018.1544360.

Tondo, L., and A. Giuffrida. 2018 Vulnerable Migrants Made Homeless After Italy Passes "Salvini Decree". *The Guardian*, December 7.

Triandafyllidou, A., and V. Gerrand. 2015. *Re-immaginare la cittadinanza: un confronto tra politici, ricercatori e scrittori*. Florence: European University Institute.

Werbner, P. 2006. Vernacular Cosmopolitanism. *Theory Culture & Society* 23 (2–3): 496–498.

Wu, Ming, and A. Mohamed Marincola. 2012. *Timira: Romanzo meticcio*. Torino: Einaudi.

Zincone, G., and M. Basili. 2013. *Eudo Citizenship Observatory*. Country Report, Italy.

Zinn, D.L. 2011. "Loud and Clear": the G2 Second Generations Network. *Journal of Modern Italian Studies* 16 (3): 373–385.

Interviews

Gerrand, V. 2019. Interview with Marilena Fabbri, Bologna, February 18.

Gerrand, V., and A. Sabrie. 2015. Interview with Kaha Mohamed Aden, European University Institute, Florence, April 28.

Visual Media

Johnson, E.M. 2017. Photograph of Magic Johnson and Samuel Jackson taken at Forte dei Marmi, Italy, August 17.

neriitaliani.blackitalians Instagram account, February 2019.

Spinazzè, P.P. (CIBO). 2018. This artist is fighting fascism with cheese. *Vice News Tonight* HBO.

CHAPTER 11

The *Habesha* Italians: The Black Mediterranean and the Second-Generation Condition

Giuseppe Grimaldi

Introduction

On October 2013, following the Lampedusa shipwreck, a group of Milanese of Ethiopian and Eritrean origins organized a commemorative march for the 368 victims. The mobilization was the first step of a wider engagement in the Milanese space about forced migrations that took place between 2013 and 2015. During this first part of the so-called "refugee crisis,"[1] hundreds of Ethiopian and especially Eritreans asylum seekers arrived every day in Milan. They gained public relevance in the Milanese discourse on forced migrations specifically as children of immigrants. In

[1] The terms "refugee crisis" as well as "refugee flow" have been critically addressed since the early 90s as Western hegemonic and de-historicized constructions. (See Malkki 1995.) All through this work, I will use these terms within a Black Mediterranean perspective in order to make sense of their hegemonic value in the present European social configuration.

G. Grimaldi (✉)
University of Verona, Frontiera Sud Aps, Italy

G. Proglio et al. (eds.), *The Black Mediterranean*, Mediterranean Perspectives, https://doi.org/10.1007/978-3-030-51391-7_11

fact, they emphasized a common *Habesha*[2] ethnicity among the asylum seekers to represent themselves as proper stakeholders to the Milanese municipality.

In contemporary Europe, links to non-European ethnic, religious or national engagement or ancestral land among people born and raised in Europe are undoubtedly increasing. The explanation of this phenomenon is often connected to a widespread paradigm. Children of immigrants, in fact, are mainly considered as subjects "in-between" two different social worlds: their parents' ancestral context and the place they were born and raised. This paradigm became so strong in both the public and academic discourse that it has even become celebrated: Levitt (2009), for example, considers in-betweenness as a constituent part of immigrants' social life and the basis of their transnational embeddedness.

> I argue that we should not dismiss outright the strong potential effect of being raised in a transnational social field. When children grow up in households and participate in organizations in which people, goods, money, ideas and practices from their parents' countries of origin circulate in and out on a regular basis, they are not only socialized into the rules and institutions of the countries where they live, but also into those of the countries from whence their families come. They acquire social contacts and skills that are useful in both settings. They master several cultural repertoires that they can selectively deploy in response to the opportunities and challenges they face. (Levitt 2009, p. 1226)

This was the perspective that oriented my analysis during my fieldwork on Italians of Ethiopian origins' "ethnic mobilization" (Grimaldi 2018b): performed *Habeshaness* seemed to be the material representation of their being "halfie" (Silverstein 2005). However, despite being very intuitive, that perspective was not convincing at all when confronted with an empirical analysis. During my ethnography, I realized that the most of the Habesha activists had almost no knowledge of their ancestral land's

[2] The term *Habesha* is a slippery concept concerning the ethnic as well as the racial and the political sphere in the Horn of Africa. It has been used to name geographical pockets of territory and people extending from the Arabian Peninsula to the furthest limits of the Horn of Africa region. The analysis of the term and of its different social and political meanings (the term is very problematic both in Ethiopia and in Eritrea) is far beyond the scope of this work. For an analysis of the term and its conjugations, see Smidt (2014), *The term Habäša Habš—An Ancient Ethnonym of the "Abyssinian" Highlanders and its interpretations and connotation.*

language, nor of the cultural habits of the asylum seekers. Most of the activities they arranged in Milan, furthermore, found the indifference of the Milanese Ethiopian and Eritrean diaspora and the overt criticism of their parents' generation. The older generations criticized especially the way the new generations built up the mobilization and their use of the term Habesha to overcome the Ethio-Eritrean national divisions. The evidence showed an incontestable reality: the nexus between children of immigrants' ethnic engagement and an assumed cultural competence related to the ancestral land is all but consistent (Grimaldi 2017, 2019). In short: many had a rather superficial idea about what is going on in their ancestral land but decided to represent themselves as part of that social world.

Children of immigrants' performances of the cultural and symbolic meanings related to their ancestral lands are growing more and more in the contemporary panorama. These re-appropriation patterns cannot be categorized as mere contingent processes. Ethnic revivals or religious fundamentalism among children of immigrants in Europe, on the contrary, question the very idea of a European citizenry. This process leaves many questions open. In this article, I will try to delineate a perspective in the analysis of the ways children of immigrants perform ancestral identification. Are we sure that ancestral identification relates to the ancestral land of children of immigrants' social lives? What if we imagined ancestral identification as a tool to focus on the place where they were born and raised?

In this work I will focus on the term Habesha and the ways its meaning has changed in the experience of Italians of Ethiopian and Eritrean origins: from an ethnonym related to the Horn of Africa to the marker of their very social condition in Italy. One of the key features of a Black Mediterranean perspective is the possibility to work on present configurations among racialized Europeans in the light of *longue durée* processes intersecting the production of a European modernity as a discrete racial unit (Danewid 2017). By focusing on the changing meanings related to the term Habesha among the Italians of Ethiopian and Eritrean origins, I trace how colonial relations between Italy and the Horn of Africa reverberated in the production of their subjectivities. The analysis of the term Habesha, in this perspective, will mirror the structures of inequality reproducing the very concept of Italianness, as well as the ways this concept acted on children of immigrants' social life. This configuration of the term Habesha as the marker of a racial, cultural and symbolic differentiation compared to the Italian context may illuminate key processes reproducing national identity in the Black Mediterranean. The analysis of children of immigrants' identification patterns reveals how colonial and postcolonial structures reproduce a race-class-gender differentiation and illuminates the relationship between racialization and the production of national identity in Europe. Drawing on the analysis of the constitutive processes reproducing Afro-European identity in the Black Mediterranean, the analysis proposes to make sense of the very concept of "second generation" as the referent of a historically, socially, and symbolically determined differential condition.

The article draws on the multi-sited longitudinal fieldwork I conducted for my PhD thesis with Italians of Ethiopian and Eritrean origins and their friends and relatives in Milan, Addis Ababa and London between 2015 and 2017. The ethnographic sources draw on extensive participant observation and a set of semi-structured interviews. The age of the people involved ranged from in their late 20s to late 30s. For privacy reasons, I changed the names and the personal information of the people I interviewed.

The Second-Generation Condition

Since its formulation in the early 1990s as a marker aimed at describing a new social category arising from the opening of new mobility patterns in South-North relations (Portes and Zhou 1993), the expression "*second generations*" has been widely criticized both as a descriptive and as an analytical tool. Many authors have emphasized the predicament related to the term both in a synchronic and a diachronic perspective: the term "second generation" does not distinguish the age of arrival of the subjects entitled to fit this social category. In addition, it lacks diachronic depth because it associates different generational experiences (Mannheim [1923] 1970) like differences between "millennials" or "baby boomers", for example. Furthermore, the term encompasses any legal status: both citizens and migrants (regular as well as irregular), in fact, can be defined as "second generation." Finally, the use of the term "second generation" hides the racial differentiation processes constituting the relation between children of immigrants and the so-called *receiving context*.

As I mentioned above, the idea of "in-betweenness" among children of immigrants, however, has entered very powerfully in the public discourse. It has even found a graphic connotation. Ambrosini (2009), for example, talks about hyphenated identities and Italians with the hyphen, as subjects in between the place where they were born or raised and their ancestral land: according to this perspective, the idea of "second generation" itself would represent a semantic referent aimed at bridging the gap between two differentially ordered social as well as symbolic spaces.

In relation to the Italian context, it seems clear that the term "second generation" certainly did not emerge as a self-imposed definition, but rather as a need for the receiving societies to label and count immigrants and their offspring (Thomassen 2014, p. 26). In this respect, Thomassen argues that the analytical representation of the so-called *second generations* as subjects in-between two cultures is based on false assumptions and that the real division lies in the discrepancy between "rhetoric and reality" in the understanding of their social lives (Thomassen 2014, p. 35). More accurately, the term has become a discursive tool fueling the Italian structural contraposition between "nationals" and "foreigners." In the present Italian scenario, the term *second generation* structurally configures a liminal space situated between the space of modern national identification and colonial otherness (Said 1978): a colonial fracture (Bancel et al. 2005; Berrocal 2010) aimed at reproducing the social, racial, and symbolic boundaries of the concept of "Italianness." According to Gaia Giuliani (2014), Italianness, in fact, can be defined as an altero-referential discourse stemming from the Italian colonial experience[3]: a process of social, racial, and symbolic differentiation from a designated "other."

Second Generations or Differential Italians?

The concept of difference sustaining the social category of "second generation" as a descriptive term in Italy needs to be considered as a central axis of enquiry rather than as the starting point of the analysis on children of immigrants' social life. Rather than describing a relation of diversity, children of immigrants' pretended in-betweenness can be framed as the result of historical relations of power aimed at reproducing hegemonic and subaltern subjects.

[3] From the immediate post-unitarian period (1870) to the Fascist racial laws (1936), the processes of social and racial differentiation represented a core feature in the processes reproducing the concept of Italianness. In this respect, drawing on Guillaumin (1995 [1972]), Giuliani (2014) underlined how the major version of Italianness derives from a system of racialization that is centered on the other: the assignment of a precise color produces the racial identity of the self (p. 573).

The construction of an internal differentiation within the modern nation-state underlines the ongoing reproduction of a colonial paradigm (Mezzadra 2008, p. 32). In the Italian case, where the *colonial subject* represented the counterpart of the production of a racially connoted and unitarian Italianness, the processes that since the end of the twentieth century reproduced an alterity within the national space (postcolonial mobility, transnational labor migrations and the new millennium forced migrations) have been framed through the same perspective. The nation-state's differential paradigm, which reproduces the isomorphic relation between citizenship, state and territorial belonging, therefore, produced second-class citizens (Balibar 2003 p. 191) who are structurally differentiated within the national body. This process does not only happen on a legal level, with the structural limitations in the formal acquisition of citizenship in the Italian legislative system.[4] Considering the Italian national paradigm where the idea of citizenship lays in the intersection of a pretended territorial, cultural, and racial continuity, even Italian citizens can be structurally differentiated if they do not meet the characteristics reproducing hegemonic Italianness. The term "second generation," emphasizes the beginning of a never-ending continuity—after the second generation, we have the third, the forth and so on. From a racial, cultural, symbolic perspective with an incommensurable otherness, the term can be framed as the material representation of these differentiation processes. Far from being a mere sociological descriptive category made up of children of immigrants, therefore, the term "second generation" needs to be explored as the referent of a historically, socially, and symbolically determined social condition. The semantic resignification of the term "second generation," I argue, needs to start from its syntactical reformulation. Through a process of adjectivization and hyphenation, it is possible to identify a proper "second-generation condition" as a space of incorporated differentiation within the modern national paradigm. The inscription of this conceptual shift within the framework of the Black Mediterranean paves the way to opening an analytical space where the past and the present structures reproducing national identification through *otherness* overlap. Colonial and postcolonial structures, transnational migrations, and the present

[4] In this perspective, the failure of the citizenship reform in Italy, the so-called *Ius soli*, is very revealing. The draft law, presented in 2013, was stalled in Parliament for the duration of the whole legislature until its final sinking in December 2017.

Mediterranean route redefine and positionally determine the ongoing reproduction of a second-generation condition.

Children of immigrants, however, incorporate the processes acting on their social condition in unsettled ways. This is the fate of the term Habesha, an ancient ethnonym related to the Horn of Africa's social space that the Milanese of Ethiopian and Eritrean origins incorporated as a form of shared identification to navigate their differential condition.

From the Horn of Africa to Italy: Habesha in the Black Mediterranean

One of the most performative values associated with the term Habesha is undoubtedly its racial value: the term Habesha, in fact, emphasize a structural difference compared with the wider sub-Saharan Africa. Abyssinia, the land they inhabit, not by chance, had been defined in the European orientalist tradition as the African anomaly and paradox (Jesman 1963). As Tibebu (1996) pointed out, Habesha were considered as "Semitic", rather than "Negroid"; "civilized" and not "barbaric"; "beautiful" and not "ugly", and so on. In the Western paradigm of knowledge, therefore, the term Habesha was explained in superlative terms because the "Negro" is considered subhuman (p. 427).

Some of the main rhetoric founding the hegemonic features of the term assumed a political continuum (Levine 2014 [1974]) from the Axumite civilization to the late nineteenth century production of modern imperial Ethiopia. Following the Wuchale[5] treaty and the making of the Italian Eritrean colony, however, there was a political fracture within the hegemonic Habesha group. The Christian, Tigrinya speaker groups inhabiting the highlands of the present Tigray region in Ethiopia and the *Kebessa* (highland) in Eritrea found themselves divided by a border and engaged in two different national paradigms (Smidt 2010). From this moment on, Italy turned out to represent a central axis of power to make sense not only of the relation between these two groups, but even of the making and unmaking of the relations redefining the meanings of the term Habesha. In fact, while Italy played an important role in the production of the two

[5] With the treaty of Wuchale (1889), the Emperor Menelik II dealt out a part of the Abyssinian territory to the Italian Kingdom. One year later, the official Eritrean colony was born.

national paradigms (and especially of Eritrean nationalism[6]), the Italian presence in the Horn of Africa, at the same time, connoted the relation between the colonizer and the colonized through the term Habesha.

In the major urban settings of both Ethiopia and Eritrea, Italians configured a colonial elite before, during and after the colonial occupation (Marchetti 2014). In this setting, the attributions connoting the term *Habesha* as the "non-Black of Africa" faded away in a racial-class relation with the Italian pre-colonial, colonial and postcolonial elite: a relation reproducing the ethnonym Habesha as the referent of Black domestic workers, laborers, concubines, and servants.

The Italian presence in the Horn of Africa massively decreased in the 1970s because of upheavals in the area, the deposition of the emperor Haile Selassie, and the instauration in 1973 of a Marxist-Leninist military junta (the Derg). The massive migrations back to Italy, however, involved even the Ethiopians and the Eritreans bounded to the former postcolonial elite. These people represented the first strong wave of postcolonial migrants and the first activators of the migration chains that, from the mid-1970s on, connoted the transnational mobility from the Horn of Africa to Italy (and Europe).[7] Once in Italy, they were involved in the same unequal structures reproducing their relations with Italians in the Horn of Africa. The Habesha migrant presence in Italy that developed from the late seventies on, therefore, configured as a proper mirror of their colonial and postcolonial condition on the opposite shore of the Mediterranean.

In the next sessions of the article, I will focus on the offspring of these Habesha migrants, and I will show how the constitutive features of the Black Mediterranean as a space reproducing the Euro-African colonial fracture impacted their social lives. Through the exploration of these structures of inequality, I argue, it is possible to make sense of the ways Italians of Ethiopia and Eritrean origins re-signified the ethnonym Habesha.

[6] On the relation between Italian colonialism and the making of modern Eritrea, see Chelati Dirar 2007.

[7] On the relevance of the Habesha transnational networks, see Grimaldi 2018b.

The Postcolonial Condition and the Second-Generation Condition: Family Resemblances

> When I arrived in Milan, I was eleven. It was the New Years Eve of the 1980. I had grown up with my aunt in Asmara and I attended the Italian Catholic school. My mother sent her money from Italy. She had been in Milan since 1974, when the Italian family she was working for moved from Asmara due to the riots. I tried to run away from school two times because I wanted to enrol with the rebels. Both of the times they sent me away because I was too young [...] My mother told me that she decided to make me join her in Milan precisely because of this reason. She was worried that the next time I tried to enrol, they would have accepted me [...] The main problem was the passport. The Derg would have never allowed me to leave the country. Fortunately, we had some contacts at the immigration office in Addis Ababa and my family managed to corrupt an officer. They had to pay 100,000 birr in order to allow me to get the passport.[8]
>
> Yonas, 48 years old. Ethiopian citizen raised in Milan.

> Yes my parents told me about being Habesha when I was a kid. I remember that almost every time I cried they used to tell me: Are you throwing a tantrum? You are Habesha.[9]
>
> Tesfay, 27 Years old. Italian citizen of Eritrean origin, born and raised in Milan.

According to its academic definition, both Tesfay and Yonas could be defined as second-generation.[10] It would be even superfluous to underline the structural differences between their lifepaths and the impossibility to frame their condition under a common lens. Tesfay is an Italian laborer in Milan. He speaks very poor Tigrigna and his last time in Eritrea was in 1997, before the Ethio-Eritrean conflict. Yonas is a shop owner in Addis Ababa. He was actively involved in the Tigrayan diasporic organization in Milan against the Derg from the mid-1980s on.

[8] Fieldnotes, Addis Ababa 02.02.2016.

[9] Fieldnotes, Milan 15.05.2015.

[10] In this respect, Rumbaut (2004) proposed the classification of 1.25–1.5 and 1.75 generations in order to distinguish between life stages and generational cohorts in the "second generation experience."

Yonas and Tesfay mirror Andall's definition of the "old" and the "new" second generations.[11] Their parents' generation moved to Milan from the beginning of the 1970s to the mid-1980s, following the trajectories uniting their postcolonial condition and the transnational upheavals in the Horn of Africa. In the coalescence between these structures and their evolution along space and time, it is possible to trace the production of a second-generation condition.

Children of Domestic Workers

Until the end of the 1980s, the Ethiopian and Eritrean presence in Italy was mainly gendered. Approximately the 85% of the Eritreans in Italy were women (Anselmi 1987), and 89% of them were employed as domestic workers in major Italian cities (Scalzo 1984). As Marchetti reported (2011, p. 111), the Italian missions and churches in Eritrea played a decisive role in the Eritrean domestic workers' migration paths. Besides their role as labor *brokers* (p. 100) for institutions and bourgeois families in Italy, they had also held a central space in the process of postcolonial Italianization in Eritrea (Chelati Dirar 2007).[12] All during the Italian colonial and postcolonial presence in Eritrea, Italian Catholic boarding schools reproduced a proper *colonial authority* (Bhabha 1994) by teaching the cultural norms and values of the colonizers to colonized subjects.[13] Italian education in Eritrea was successful as a tactical tool to gain access to the hegemonic Italian socioeconomic context (Marchetti 2011, p. 79).

Domestic workers' children, who grew up in highly Italianized contexts, were among the main recipients of this postcolonial religious education. Once these children moved to Italy with their mothers (as in Yonas' case), the main religious denominations operating in Ethiopia and Eritrea

[11] Focusing on the Italians of Ethiopian and Eritrean origins, Andall (2002) underlined the differences between those who were born or raised in Italy in the 1970s and the ones who grew up in Italy from the mid-80s onwards (pp. 392–394). She distinguished, in this respect, between older and younger second generations (p. 390) as two incommensurable social groups, as they grow up in two structurally different settings. Compared to the invisibility of the older second generations, Andall states, the younger group grew up in a social and political climate where the issue of immigration and the presence of the children of immigrants had already acquired some prominence (p. 390).

[12] For a deeper account of the present function of the postcolonial Italian education system in Ethiopia, see Chap. 6.

[13] For detailed accounts of the Italian boarding schools in Asmara, see Ben-Ghiat and Fuller 2005; Marchetti 2011, Ghidei Biidu and Marchetti 2010.

(Capuchins, Combonians, Consolata and Pime) (Cataldi 2015) reproduced their postcolonial function in Italy. From the late 1970s on, most of the domestic workers' offspring in Milan, whether they were born in Italy or not, were placed all around Lombardy in religious boarding schools that originated from the convergence between religious organizations and voluntary sector involvement.[14] Before the transition of domestic work from residential to hourly wage labor, a phenomenon that grew increasingly from the mid-80s on (Mezzetti and Stocchiero 2005, p. 20), residential boarding schools (based on the same model as the postcolonial schools in Ethiopia and Eritrea) represented one of the main sources to face the "first" second generations' childcare issues (Martignoni 2016). Children of immigrants' formation in the boarding schools had been strongly marked by racial differentiation. The colonial fracture reproducing hegemonic representations of Italianness mirrored in the boarding schools' organizational patterns and provided ways to manage the Italians of Ethiopian and Eritrean origins' presence.

Alex, an Italian of Ethiopian origin in his 40s, clearly exposes the differential treatments that he and his friends received during their youth in the boarding school. Alex told me about the boarding schools out of Milan and the difficulties he found in living outside his familiar context. Boarding schools were all around Lombardy (he talked about *collegi* in both Monza and in Bergamo which each housed more than fifty Ethiopians and Eritreans). He was allowed to see his mother once a week. Most of the times at the boarding school, he was only with Ethiopians and Eritreans, despite their different ages or times of arrival. Alex justified the division between Habesha and Italians in the boarding schools because of the shared Tigrigna language and the fact of feeling stronger in a group.[15]

In the experience of the Italians of Ethiopian and Eritrean origins at boarding schools, it is possible to trace the forerunner of a socially determined second-generation condition. Boarding schools' organization had a relation of continuity with the postcolonial patterns, reproducing the relation between Italy and the Horn of Africa. The mechanisms of concentration of the Italians of Ethiopian and Eritrean origins and the processes

[14] Children of Ethiopian and Eritrean domestic workers were sent to mixed boarding schools or residential structures that had been founded to face their presence. One of the most important moments in the collective memory of both Ethiopians and Eritreans in Italy was the *Fondazione Pernigotti* that in 1980 opened the *Collegio San Marco* with the aim of sustaining the children of Ethiopian and Eritrean immigrants (Cologna Breveglieri 2003).

[15] Fieldnotes, Milan, 15.05.2016

of separation from the wider Italian audience produced a social space of differentiation. In this perspective, Italians of Ethiopian and Eritrean origins discovered themselves as a social group: a group produced through the differential lens of the hegemonic representations of Italianness in a structural continuity with their mothers' racialized, gendered, and class-oriented working activities.

Downward Assimilation: Why Should a Milanese Habesha Speak Neapolitan?

The condition of structural invisibility of the first-generation migrants from Ethiopia and Eritrea until the mid-80s and their institutional differentiation in terms of access to the welfare state represented a primary problem in the ways they built their social life in Milan. In 1978, only 10–15% of Eritreans and Ethiopians could afford a private flat. The others shared rooms with compatriots or relied on boarding houses, religious communities, and public dormitories (Martignoni 2016, p. 130). A very widespread phenomenon was, furthermore, the irregular occupation of empty houses. In 1982, around 35–40% of the Eritreans in Milan were squatters (Ambroso 1987), helped by the leftist movements in Milan.[16] Ethiopian and Eritrean migrant housing followed the wake of the Milanese proletarian class, mainly southern Italian migrants who, from the 50s onwards, kept moving to northern Italy (Boffi et al. 1973). Similarly to southern Italian immigrants, the phenomena of squatting or renting crumbling buildings in the city center provided an immediate solution to their housing problems. Assigned council housing, which from the mid-80s on started to include Ethiopians and Eritreans, generated dispersion in the Milanese suburban areas and expelled lower classes by means of the ongoing pattern of the gentrification of the city center.[17]

This process of expulsion reinforced the isomorphic relation between migration status and social class that had been initiated during the 1980s. Milanese suburbs turned out to represent specific social units disconnected

[16] The 1983 Resolution of Milano council allowed migrants to be eligible for council housing. In this perspective, squatting turned out to be a tactic of visibility in order to have a council house assigned (Martignoni 2016 p. 131).

[17] Ethiopians and Eritreans mainly concentered in the areas around the city centers built from the 1950s on. The main concentrations can be found in the council houses of the neighborhoods of Barona (north of Milan), Maciachini (east of Milan), Corvetto (south of Milan), and especially San Siro (west of Milan).

from the wider Milanese social fabric, structurally lacking social services and facilities and mainly inhabited by southern Italian migrants. The context where the most of the children of Ethiopian and Eritrean migrants were born and raised, therefore, configured as a ghetto, as a "historically determinate, spatially-based concatenation of mechanisms of ethnoracial closure and control" (Wacquant 1997, p. 343). During my fieldwork, I had the possibility to notice the widespread connection between the ways that the children of Ethiopian and Eritrean migrants represented themselves and the social setting in which they were born and raised. This process usually emerged at the beginning of a social relation with Italians of Ethiopian or Eritrean origins as soon as they heard my accent connotating my southern Italian origins. It was quite common for me to receive questions about my hometown. They often mentioned their closest Neapolitan childhood friends, or even started to talk with a Neapolitan accent. There was a widespread use of the derogatory term *Terrone* (a term used to insult southern Italians) as a source of recognition. As Magda, an Italian of Eritrean origin in her 30s told me:

> I do not want you to misunderstand me when I use the term *Terrone*. This is not an offence to me. We are very similar in culture. I grew up with my neighbours from Aversa.[18] I think we Habesha are Terroni as well.[19]

My geographic origins, paradoxically and unexpectedly, resonated with their social condition (Wikan 2013 [1992]) and reduced social distance. This process is strongly indicative of the relation Italians of Ethiopian and Eritrean origins created with the Milanese social space. Structurally excluded from the hegemonic representations of the city, they grew up in the marginal context of the Milanese suburbs, associating with the structural condition of marginalization connotating the historicized southern Italian underclass living in Milan. As Gupta and Ferguson (1997) stated, place production, far from being based on the recognition of a cultural similarity or a social continuity, is the effect of a classificatory system founded on various forms of exclusion and of the construction of otherness that defines personal and collective identifications (p. 13). The mechanisms of exclusion aimed at reproducing the Milanese hegemonic representations of the city produced a condition of structural subalternity, from whence new social configurations and affiliations arose.

[18] A small town close to Napoli.

[19] Fieldnotes, Milan, 13.04.2015.

Southern Italian housewives, for example, were a source of crucial importance in the early stage of children of Ethiopian and Eritrean origins' lives, providing a solid and cheaper alternative for care to Catholic boarding schools. Those who were born from the mid-1980s to the beginning of the 1990s (meeting the Andall's definition of the "new" second generation) were mainly raised in southern Italian families and socialized in the southern Italy's migrant setting. The structural lack of policies to face the coming age of a multicultural society in Italy led to informal practices of assimilation where marginalized Italian groups served as a vector of Italianness. This process was highly performative on Italians of Ethiopian and Eritrean origins' social positioning. First of all, it acted on their linguistic competence: rather than Tigrigna, they were exposed to the Italian language or to southern dialects in their early life stages. As Adiam, an Italian of Eritrean origin in her 30s, stated:

> Our linguistic competence is mainly due to our summer holidays in Asmara. The more you were there as a baby the more you learned Tigrigna. I think very few of us learned Tigrigna in Milan as children. On the contrary, in Milan we forgot the little Tigrigna we learned in Asmara.[20]

Boarding schools and housing issues, two of the main narratives forging the Ethiopian and Eritrean presence in Milan, played a decisive role in the making of the second-generation condition as a discursive space of structural differentiation from the hegemonic paradigms sustaining Italianness. The assimilation to a socially connotated setting, especially for males, powerfully worked on their social positioning.

Racial Differentiation, Gender, Class: The Making of the Habesha Second-Generation Condition

Ragazzi di Zona: The Class Connotation of the Habesha Identification

> Those people are still doing the same things they did when they were sixteen. They remained *ragazzi di zona*. They just think about smoking joints, having holidays, buying a car, spending their days doing nothing. They say

[20] Fieldnotes, London, 14.06.2016.

> I am a betrayer of Eritrea but they do not even know what Eritrea is. It is crazy if I think I was like that until some years ago.[21]

Tesfay, an Italian of Eritrean origins, pronounced these words while talking with me about the threats he received from some of his old friends because of his involvement with refugee reception in Milan. Tesfay explained to me the motivations that led them to classify him as a betrayer by blaming their condition on being *ragazzi di zona*, an expression that it is possible to translate as ghetissards.[22] In his opinion, their restricted view of the world led them to reproduce a fake and uncritical nationalism without any real consciousness of the Eritrean complexity. I will explore the ways nationalist politics acted on the production of a second-generation condition in this section, considering the ways the Italian males of Ethiopian and Eritrean origins grew up as *ragazzi di zona* and the social meaning it entailed.

Talking about the Black-White structural asymmetric relation, Fanon (2008 [1952]) underlines how a Black man in a white society is required to fit a series of social expectations to perform his condition. Those expectations are structurally incorporated into the process of individualization where racial attribution coalesce with class and gender. Italian men of Ethiopian and Eritrean origins born and raised as *Blacks* in Milan had to confront themselves with these processes since their transition from childhood to adolescence.

In the lines below, Johnny talks about his childhood in Milan. Johnny's words serve as a starting point to make sense of the construction of a racially connoted masculinity as a constitutive feature of the "second generation".

> One of my teachers once told me "can you please move out of [away from] the blackboard? It is of your same colour and I can't see you." It was a matter of survival at school. You had to be meaner than the others. If not, you were dead.[23]

The explicit link between racial attributions and behavioral patterns that Johnny underlined emphasizes the ways his social construction arose

[21] Fieldnotes, Milan, 15.03.2015.

[22] Milan is divided into administrative areas. *Ragazzo di zona* is a derogatory term to talk about people who organize their everyday life in the social space of the ghetto.

[23] Fieldnotes, Addis Ababa, 21.01.2016.

as a response to a condition of structural differentiation. From his adolescence onwards, he had to confront his structural condition of marginalization (on an economic, social, and symbolic level) and the social attributions he had to confront with as a Black Italian male within the Milanese setting. This connection represented a central issue in the individuation process of Italians of Ethiopian and Eritrean origins. The phenomenon of being asked for documents by the police on the streets or being asked for drugs by same-age White Italians came to be an integral part of their relationship with the city's social fabric. The suburbs where they had been raised, in these respects, represented powerful places of identification as well as arenas where it was possible to put on stage their social condition. Structurally disconnected from the Milanese hegemonic space, Milanese suburbs offered them the possibility to make sense of their differential condition. The ghetto street life where legality and illegality mingle turned in a space of social recognition. If from one side it alienated them from the Milanese public arena, on the other it allowed them to produce a site of identification out of the structural differentiation they were confronted with.

The expression *ragazzi di zona*, from this perspective, represents the marker of a downward assimilation process that forged Italian of Ethiopian and Eritrean origins' social behavior as well as their social positioning compared to the wider Milanese context. As I will show in the next paragraphs, this model entangles with their social identification as Habesha.

The Habesha Dark Side: Second-Generation Condition and Gender Divide

> Do you know about the Habesha dark side? Or rather, the dark side of the Habesha men? At my graduation party, the father of one of my friends complimented me. His daughter would reach her graduation soon as well. He told me: You girls are the pride of our community. On the contrary, I do not know what these guys have.[24]

These are the words of Serena, an Italian of Eritrean origin living in London. She introduced the issue of gender differentiation among the Italian of Eritrean origins by reporting the sentence of her father's friend. These words are very indicative for delineating some crucial traits of the second-generation Habesha condition. On the one hand, they make

[24] Fieldnotes, London, 21.06.2016.

visible the gender line dividing the Italians of Ethiopian and Eritrean origins' social experience in the Milanese context. On the other hand, they emphasize the structural distance between the youth behavioral patterns and their parents' expectations.

As Thomassen (2014) stated, there are specific male problems related to the integration of second generations in Italy. Even if men are more "free" (p. 38) on paper, it is almost the norm that young women are usually more educated than young men in Ethiopian- or Eritrean-Milanese families. All throughout our interview, Serena talked about the strong division between Italian men and women of Eritrean origins. She emphasized how criminality, drug and alcohol consumption, and even suicides have always been a constant preoccupation in Milanese Habesha families.

The gender divide running across the Ethiopian and Eritrean diasporic social space is relevant to the ways the Italians of Ethiopian and Eritrean origins socially represent themselves, and it emerged several times during the ethnographic data collection. In one of my meetings with Serena, we were in a coffeehouse in Brixton with some of her Italian friends. The coffeehouse was in front of a Habesha club. Some elderly people were sitting out of the club minding their own business. While we were having our drink, Serena jokingly asked us to finish soon and to move somewhere else since she wanted to smoke a cigarette. When I asked her why we had to move, she indicated the Habesha club. I asked her if she knew some of the elderly people sitting outside the club. She said that it was not a matter of hiding her smoking. In her opinion, it would have been disrespectful to smoke in front of old people and she would have created an unpleasant situation if they saw an Habesha woman smoking.[25]

The opposing relation between the "control" of women and the "freeness" of men as a constituent feature of Habeshaness emerged even in the explicit representations of male subjects: a culturalist perspective is often used as explanation of gender division and defined behavioral patterns. Miki, an Italian of Eritrean origin in his late 20s I met in London, explained this dynamic:

> We should be erecting a statue to our mothers. Our fathers, on the contrary, are jerks. They like to go roaming and to spend their times in clubs. Divorce is just normal. What can our mothers do when they see their husbands coming back home late at night drunk almost every day? That is why they would

[25] Fieldnotes, 31.05.2016.

> accept me to go around all day long. But they would never allow my sister to do the same.[26]

The intergenerational continuity Miki represented, however, is far from the acritical reproduction of crystallized social patterns related to the ancestral land and overlapping time and space. On the contrary, the self-attribution of the social patterns related to the ancestral land can be considered as a proper self-fulfilling prophecy. The incorporation of a continuist perspective hides the ongoing structures of inequality reproducing the so-called Ethiopian and Eritrean community in Milan as a discrete unit regardless of the social, legal, or symbolic status of the people involved.

Therefore, what is possible to define as the Habesha model of masculinity, what Serena defined as *the Habesha dark side*, far from drawing on their ancestral social habits, developed through the modes of reproduction of Milanese marginalized groups. The result was the production of an undifferentiated marginalized social unit stemming from the process of peripherization aimed at reproducing the center (Callari Galli 2000). Especially among men, therefore, their slang and way of representing themselves were strongly connected to the Milanese underworld. Downward assimilation, together with the structural differentiation they experienced as racialized Italians forged a culturally defined model of masculinity among the Italians of Ethiopian and Eritrean origins. Those structures of inequality had been incomparably more effective than their ancestral belonging in the ways they were situated within the Italian setting.

The lack of social mobility, the racializing paradigm they were subjected to, and their structural exclusion from the hegemonic representation of the Italian social space produced a terrifying immobility across generations. This condition is clearly expressed in the following sentences of Joe, an Italian of Eritrean origin in his mid-30s, who moved to London because of the lack of opportunities of the Italian context. Talking about his youth in Milan he said:

> Our mothers used to clean up the asses of rich old people. We sold drugs to their sons. We had good contacts to find it. And all of us started from here, in a society that have never accepted us. So fuck school. I sold weed and

[26] Fieldnotes, 06.07.2016.

coke. Some of us went too far with it. But it was a way to show we were something.[27]

Resignifications: Appropriating the Term Habesha as the Marker of the Second-Generation Condition

Until my ninth birthday there were at least fifty guests at home every year. From 1998 on, more than the half of them stopped coming.[28]

These words were pronounced by Jimmy, an Italian of Ethiopian origins who was born in Milan in 1989. Jimmy told me about the birthday parties of his childhood to emphasize the ways social relations changed in the Ethiopian and Eritrean diasporic social space from 1998 on, when the Ethio-Eritrean conflict started. The political upheavals in the region that turned the relations between the two states from allies to enemies (Tronvoll 2009) represented not only a watershed in the recent history of the Horn of Africa (Hepner and O'Kane 2009), but heavily also reverberated on diasporic configurations.

Since the end of the 1990s, a big shift in the communitarian structures sustaining the Milanese diasporic social space took place. The Milanese shared social structures that from the late-1970s on represented the reference point in the Ethiopian and Eritrean diasporic social life gradually lost their attractivity in favor of nationalist structures.[29] The shift from a communitarian to a nationalist configuration, far from being related to the mere political sphere, entailed social, religious, and symbolic structures mirroring the national contrasts in diaspora. As Miriam, an Italian of Ethiopian father and Eritrean mother told me:

I have been lucky. You cannot imagine how many families split up because of the war. People who divorced, family members that stopped talking to each other. It was such a mess.[30]

The most significative example of the affirmation of a nationalist configuration was exemplified in the fate of the oratory of Via Kramer (the

[27] Fieldnotes, London, 16.07.2016.

[28] Fieldnotes, Milan, 11.07.2015.

[29] In July 2018, a peace deal between Ethiopia and Eritrea entered into force after 20 years of war. Its effects are yet to be explored both in the homeland and in the diaspora.

[30] Fieldnotes, Milan, 11.02.2015.

oratory of the main Ethio-Eritrean church in Milan), where young Ethiopian and Eritrean used to meet from the mid-80s to the mid-90s.

Via Kramer: Habesha Spaces and the Second-Generation Condition

Until the mid-90s, the oratory of Via Kramer represented a gathering space where different generations of Italians of Ethiopian and Eritrean origins creatively elaborated their positioning in the Milanese context. Via Kramer represented a powerful place of hybridization: a place of intergenerational continuity where, close to the ancestral land flows, meanings and practices, new models of identification arose. Those models were all but in continuity with ancestral lands' modes of self-reproduction but strongly connected to the Milanese social space. Via Kramer oratory, shortly, represented a space where they could translate their differential Italianness (in gender, class, and racial terms) into self-valuing attributes. The so-called "first" second generations (those who were born or raised in Italy in the early 1970s) found in Via Kramer a site of identification not hierarchically ordered. Via Kramer oratory exercised a centripetal force on the Italians of Ethiopian and Eritrean origins. It was a space of social recognition aggregating them from the disjointed Milanese suburbs, a place to practice their differential condition as a self-valuing attribute. From the mid-80s to the mid-90s, Via Kramer represented one of the main hip-hop cultural places in Milan, and a source of social legitimation to act in the upcoming Milanese Black global culture. The downward assimilation and the behavioral patterns it entailed, the racial paradigm that children of immigrants had to confront, and the gender lines dividing the experience of being considered Black in a White society coalesced in Via Kramer oratory in a performative model. The everyday practice of the Via Kramer oratory produced the first "Black Italian" social group in Milan: a group that found its social legitimation in the public space as the representatives of a Blackness in the 80s and 90s Black-oriented club parties in Milan. The term Habesha, in this perspective, showed its nature of liminoid space of identification (Turner 1974), expressing in leisure times practices and subversions of the structural relations reproducing their differential Italianness.

As Miki, an Italian of Eritrean origin told me, remembering about his younghood in the first 90s:

> We brought hip-hop to Milan. Any time we went in one of the clubs the owner was so happy. We were more than fifty every time. We got free drinks, free entries and the VIP area of the club reserved [for us].[31]

In this perspective, being Habesha was a way of being in the Milanese social space. The focal force of the Via Kramer oratory allowed Italians of Ethiopian and Eritrean origins to come together in the disconnected space of the Milanese suburbs and to identify themselves under a common condition.

With the growing hostility between the two countries, cohabitation became more and more complex. The communitarian function of the oratory as a bridge between the diaspora and the Milanese social space started to decrease.[32]

When Diaspora Breaks Down, the Second-Generation Condition Does Not

The rise of nationalist configurations and the fragmentation of the diasporic identification, however, did not change children of immigrants' position within the Italian space. Downward assimilation, racialization, and the structural problems it implied remained untouched after the Ethiopian and Eritrean nationalist shift; the same went for the necessity to deal with their condition. The disappearance of the diasporic structures (such as the Via Kramer oratory), therefore, did not deprive the "second" second generations of the possibility to socially situate themselves as a social group in the everyday Milanese social space.

The main difference between the "first" and "second" cohort of Italians of Ethiopian and Eritrean origins concerned the ways of reproducing their social condition. Deprived of institutionalized gathering spaces where to reproduce as a social group, young Italians of Ethiopian and Eritrean origins started to find legitimation in the public spaces of the city reproducing the cultural contents of their differential condition. In the adolescence of most of the "new second generations," the informal hip-hop meeting places of Milan (above all *il Muretto* and *Parco Sempione*) represented powerful sites of recognition. Similarly to the Via Kramer oratory, these

[31] Fieldnotes, London. 13.08.2016.

[32] With the turn of the new millennium, the oratory lost its relevance until its definitive closure. Nowadays, it has been replaced by a museum of the Capuchin priests.

places had a powerful centripetal force. Different from the diasporic structures, however, these places reunited people from all over Milan of different ethnic and social class backgrounds. In this diverse social space where the only bond was a shared cultural attitude, the everyday life of the Italians of Ethiopian and Eritrean origins was made of convergences but also dissonances with the general users of these places. Detached from the Ethiopian and Eritrean diasporic world and used as a site of symbolic recognition, the term Habesha and its cultural contents arose as the descriptor of a second-generation condition.

The Muretto: Habesha as an Italian Phenomenon

> I grew up in there but I had no contacts with the people you were talking about. The Eritreans, and the people of African origins generally had few contacts with the Milanese hip-hop scene. [...] It was not about racism. It was just because few of them were interested in making hip-hop, compared with their cultural background. [...] You should imagine them like in a parallel dimension compared to the Milanese hip-hop scene. While the Milanese hip-hop was taking a definite path, they elaborated their own models, even if we all attended the same places or the same clubs. [...] Hip-Hop is based on creativity. You have to do, not just to listen. They have always listened to a lot of music. And, nowadays, many of them are great DJs. But they have always made very little music. In this way, it is possible to understand the division of the clubs. In those where you could make music, and not just listen to music, it was very rare to see them.[33]

Chiara, an Italian White woman in her mid-30s, talked about her adolescence in the *Muretto*, the Hip-hop reference space of Milan. Her words depict a strongly fragmented space and a division between the autochthons and the Italians of migrant origins. She explicitly underlined the existence of a parallel dimension producing two different fruitions of the hip-hop places of the city. The division she underlined was confirmed by Joe, a Milanese hip-hop MC of Eritrean origins whom I met in London:

> Most of us had no interest in making music. Habesha were the kings of clubbing. We made the real club nights in Milan. We did not mix with

[33] Facebook conversation. Addis Ababa, 12.02.2016.

> Whites. We were in two different worlds. They made the hip-hop. We were the kings of clubbing. Everywhere we were used to go, we were the kings.[34]

The split within a socially shared cultural framework can be placed exactly within the intersection of racial and class structures reproducing the Italians of Ethiopian and Eritrean origins condition. The Muretto, a place in Piazza San Babila, one of the most central squares of the city, was a part of the square where people could meet and practice hip-hop disciplines.[35] The Muretto had its acme between the mid-80s and the early 90s, and its main habitués were generally middle-class young White Milanese. Despite different waves during the 90s,[36] the class configuration of the place remained the same until the late-90s, when the Italians of Ethiopian and Eritrean origin started to attend daily. The difference in terms of social, economic, and symbolic structures between the two groups resulted in the production of a fracture in the configuration of the Muretto. This way, a peculiar use of the space among the Italians of Ethiopian and Eritrean origins emerged. As a socially oriented place within the Milanese panorama, the Muretto worked as a site of recognition to reproduce their differential Italianness as a self-valuing attribution. Their class and racial differentiation represented the basis of their legitimacy in the Milanese hip-hop scene. Walking the paths of the "first" second generations, that, from the 1980s on, found in the Milanese Black musical scene a source of social recognition, the "second" second generations entered the reference space of the Milanese hip-hop by following the same cultural contents. Rather than encapsulating models of cultural reproduction in the Muretto, Italians of Ethiopian and Eritrean origins created a parallel site within the same cultural container: a parallel site based on their differential Italianness. The Muretto, under this perspective, represented a site that made the "second" second generations able to daily reproduce as a social group. The differences in the fruition of the place produced even different identification within the hip-hop social space. Similarly to the "first" second generations, their self-ascribed Blackness turned out to be a decisive element in the ways they positioned in the Muretto hip-hop scene.

[34] Fieldnotes, London, 16.07.2016.

[35] These disciplines include MCing (making music), DJing (creating music), Breaking (dancing Break dance), Writing (making aerosol art or graffiti).

[36] The place rose and declined according with the waves linked to the Milanese hip-hop movement but especially because of the structural interventions in the square (urban renewal) encouraging or limiting the Muretto as a gathering space.

Transnational mediascapes (Appadurai 1996), mainly connected to the US hip-hop, worked in the production of a social and symbolic elsewhere compared to the Milanese social space. The isomorphism between children of immigrants' self-representations and the imaginaries of the Black American hip-hop culture, especially for males, worked as a site of translation of their differential condition. The classist and racial differentiation they experienced in the Milanese suburbs had their declination in the hegemonic Black American hip-hop symbols. Social and behavioral patterns, and even colloquial language itself, were strongly oriented from these flows. The most evident phenomenon was the naming process. Almost all of the people I met in Milan during their adolescence gained their own social name. In most of the cases, the name resulted from an Americanization of their given name (Johannes turned in Johnny, Josef turned in Jay, etc.). In some other cases, they totally changed their given names and they self-identified with nicknames borrowed from American popular culture. The process of name-changing, a central dynamic in self-representation practices of identity (Rymes 1996), is strongly indicative of how the production of a second-generation condition, rather than following the homeland dynamics, is strongly oriented from the differentiation patterns they experience in Italy.

Habesha Party: Celebrating the Second-Generation Condition

This social identification was (and still it is) performatively celebrated: in all of the significant places of the Ethiopian and Eritrean diaspora, it is quite common to regularly celebrate Habesha parties, where both Ethiopians and Eritreans gather and party all night long. The contents of the party strongly differ from place to place, as well as depending on the organizers. In Milan, Habesha parties differs in their contents and modalities, whether they are organized by diaspora people or by Italians of Ethiopian or Eritrean origins. Second-generation Habesha parties are proper club hip-hop nights that alternate hip-hop songs with songs from Ethiopia or Eritrea.

The vehicular language is Italian rather than Tigrigna, the dress code is predominantly hip-hop oriented. These parties work as proper sites for reproducing an identification contrasting the outer Milanese space. Johnny, an Italian of Ethiopian origins in his late 20s, very well expresses the social function of these parties. In one of these parties, he experienced for his first time the Habesha setting:

> The first time I joined an Habesha party I thought my friends brought me among refugees. I had never seen all those Black people together. I couldn't even imagine there were so many in Milan. I grew up as the only Black among my friends. I was the only Black in my classroom. Since that time, I understood that I wasted a lot of time. Everything was easier. It was easier with girls. No one asked me stupid questions. [...] For the first time I felt powerful and legitimated.[37]

Conclusions: The Salience of the Second-Generation Condition in the Black Mediterranean

Throughout this article, the term Habesha has been framed as a racial construct, an ethnic device, a national paradigm and a source of social recognition according to the ways that children of immigrants positionally navigate their social condition. In all of these conjugations, the adherence of the term Habesha to ancestral land dynamics or within the diaspora social space itself has never been at stake. Second generations' Habesha identification emerged as a group-making project (Brubaker 2004): the effectiveness of this project, far from being related to an ancestral ascription, lays precisely in the entanglement within the classifications they are structurally confronted with.

The Black Mediterranean represents a powerful analytical umbrella to work the effects of these classifications in a time-space wide perspective. The analysis of the second-generation condition in the Black Mediterranean, this way, may open interesting analytical paths in the exploration of the concept of intergenerational continuity itself under a different light.

The case study I reported clearly shows that the structures of inequality acting on first-generation migrants in Italy, powerfully concur in the definition of a second-generation condition under an economic, symbolic, and cultural point of view. The socio-racial structures reproducing Italian national identity since colonial times are still the major driving forces in the production of black bodies as external from the national body.

These structures clearly represent proper standing points in the definition of the Italian national identity. The traumatic effects of the intergenerational nature of children of immigrants' condition, in this perspective, could become apparent in the coming decades.

[37] Fieldnotes, Milan, 15.09.2015.

A new generation of Italians raised from parents who arrived in Italy across the Mediterranean route is growing up. These people will carry on the frontier with them (Rastello 2014) and will incorporate the mechanisms reproducing "fortress Europe" as one of the main legacies of their parents' generation.

The Black Mediterranean, in this perspective, can be a vector of recognition of the past and present processes of European exploitation, violence and necropolitics toward Africa as constitutive of the contemporary Afro-European identity. These phenomena, I strongly believe, will be the touchstones from which the very idea of national and European identity (if ever existed) will be confronted in the coming years.

References

Ambrosini, M. 2009. Italiani con il trattino: identità e integrazione tra i figli degli immigrati. In *Seconde generazioni e riuscita scolastica. Il progetto SeiPiù*, ed. L. Tieghi and M. Ognisanti. Milano: Franco Angeli.

Ambroso, G.C. 1987. *'Urban Ethnicity': Culture and Politics Among Eritrean and Egyptian Migrants in Milan*. PhD Dissertation, University College of London.

Andall, J. 2002. Second-Generation Attitude? African-Italians in Milan. *Journal of Ethnic and Migration Studies* 28 (3): 389–407.

Anselmi, A. 1987. La comunità eritrea. *Dossier Europa emigrazione* 12 (4): 14–16.

Appadurai, A. 1996. *Modernity Al Large: Cultural Dimensions of Globalization*. Minneapolis: University of Minnesota Press.

Balibar, E. 2003. *We, the People of Europe? Reflections on Transnational Citizenship*. Princeton/Oxford: Princeton University Press.

Bancel, N., P. Blanchard, and S. Lemaire. 2005. Introduction. La fracture coloniale: une crise française. In *La fracture coloniale La société française au prisme de l'héritage colonial*, ed. P. Blanchard, N. Bancel, and S. Lemaire, 9–30. Paris: La Découverte.

Ben-Ghiat, R., and M. Fuller. 2005. Introduction. In *Italian Colonialism*, ed. R. Ben-Ghiat and M. Fuller. Basingstoke: Palgrave Macmillan.

Berrocal, E.G. 2010. Building Italianness Through the Logic of the 'Other in Us' and the 'Self in the Other': An Anti-nationalist Approach to the Italian Debate on a New Citizenship Law. *Bulletin of Italian Politics* 2 (1): 69–90.

Bhabha, H. 1994. *The Location of Culture*. London: Routledge.

Boffi, M., S. Confini, A. Giansanti, and E. Mingione. 1973. *Città e conflitto sociale*. Milano: Feltrinelli.

Brubaker, R. 2004. *Ethnicity Without Groups*. Harvard: Harvard University Press.

Callari Galli, M. 2000. *Antropologia per insegnare: teorie e pratiche dell'analisi culturale*. Torino: Pearson Italia Spa.

Cataldi, A. 2015. *Le missioni cattoliche italiane nelle colonie d'Etiopia e d'Eritrea*. Lecce: Grifo.

Chelati Dirar, U. 2007. Colonialism and the Construction of National Identities: The Case of Eritrea. *Journal of Eastern African Studies* 1 (2): 256–276.

Cologna, D., and L. Breveglieri. 2003. *I Figli dell'immigrazione: ricerca sull'integrazione dei giovani immigrati a Milano*. Vol. 150. Milano: Franco Angeli.

Danewid, I. 2017. White Innocence in the Black Mediterranean: Hospitality and the Erasure of History. *Third World Quarterly* 38: 1–16.

Fanon, F. 2008 [1952]. *Black Skin, White Masks*. New York: Grove Press.

Ghidei Biidu, D., and S. Marchetti. 2010. Eritrean Memories of the Postcolonial Period: Ambivalence and Mimicry in Italian Schools in Asmara. In *National Belongings: Hybridity in Italian Colonial and Postcolonial Cultures*, ed. J. Andall and D. Duncan. Oxford/Bern: Peter Lang.

Giuliani, G. 2014. L'Italiano Negro: The Politics of Colour in Early Twentieth-Century Italy. *Interventions* 16 (4): 572–587.

Grimaldi, G. 2017. Structures of Return Between Italy and Ethiopia: Mobility of the Second Generations to the Ancestral Land as a Self-Fulfilling Prophecy. *Africa e Mediterraneo* 26 (86): 54–58.

———. 2018a. *In search of Italianness. An Ethnography of the Second-Generation Condition in a Mobility Perspective*. PhD Dissertation, University of Milano Biccoca.

———. 2018b. Family First: Diaspora Houses in Mekele from Infrastructures of Return to Infrastructures of Mobility. In M. Belloni, S. Bruzzi, and V. Fusari, *Mobility and Affective Relationship*, Special Issue, Studi Emigrazione, International Journal of Migration Studies, 212: 575–592.

———. 2019. The Black Mediterranean: Liminality and the Reconfiguration of Afroeuropeanness. In A. Rastas and K. Nikunen, *Contemporary African and Black Diasporic Spaces in Europe*, Open Cultural Studies. pp. 414–427.

Guillaumin, C. 1995 [1972]. The Specific Characteristics of Racist Ideology. In *Racism, Sexism, Power and Ideology*, ed. C. Guillaumin, 29–60. London: Routledge.

Gupta, A., and J. Ferguson. 1997. Culture, Power, Place: Ethnography at the End of an Era. In *Culture, Power, Place: Explorations in Critical Anthropology*, ed. A. Gupta and J. Ferguson, 1–29. Durham: Duke University Press.

Hepner, T., and D. O'Kane. 2009. Introduction: Biopolitics, Militarism, and Development in Contemporary Eritrea. In *Biopolitics, Militarism, and Development: Eritrea in the Twenty-First Century*, ed. D. O'Kane and T.R. Hepner, ix–xxxvii. New York: Berghahn Books.

Jesman, C. 1963. *The Ethiopian paradox*. London, New York, Nairobi: Oxford University press.

Levine, D.N. 2014 [1974]. *Greater Ethiopia: The Evolution of a Multiethnic Society*. Chicago: University of Chicago Press.

Levitt, P. 2009. Roots and Routes: Understanding the Lives of the Second Generation Transnationally. *Journal of Ethnic and Migration Studies* 35 (7): 1225–1242.

Malkki, L.H. 1995. Refugees and Exile: From "Refugee Studies" to the National Order of Things. *Annual Review of Anthropology* 24 (1): 495–523.

Mannheim, K. 1970 [1923]. The Problem of Generations. *Psychoanalytic Review* 57 (3): 378–403.

Marchetti, S. 2011. *Le ragazze di Asmara: lavoro domestico e migrazione postcoloniale*. Roma: Ediesse Editore.

Marchetti, S. 2014. *Black girls: migrant domestic workers and colonial legacies.* Chicago: Brill.

Martignoni, M. 2016. *Postcolonial Organising: An Oral History of the Eritrean Community in Milan*. PhD Dissertation, University of Leicester.

Mezzadra, S. 2008. *La condizione postcoloniale: storia e politica nel presente globale.* Verona: Ombre Corte.

Mezzetti, P., and A. Stocchiero. 2005. *Transnazionalismo e catene migratorie tra contesti locali*. Cespi Working Papers, n. 16, Rome.

Portes, A., and M. Zhou. 1993. The New Second Generation: Segmented Assimilation and Its Variants. *The Annals of the American Academy of Political and Social Science* 530: 74–96.

Rastello, L. 2014. *La frontiera addosso: Così si deportano i diritti umani.* Bari: Laterza.

Rumbaut, R.G. 2004. Ages, Life Stages, and Generational Cohorts: Decomposing the Immigrant First and Second Generations in the United States. *International Migration Review* 38 (3): 1160–1205.

Rymes, B. 1996. Naming as Social Practice: The Case of Little Creeper from Diamond Street. *Language in Society* 25 (2): 237–260.

Said, E. 1978. *Orientalism*. New York: Pantheon Books.

Scalzo, F. 1984. Stranieri in Italia. La comunità eritrea e marocchina nell'area romana attraverso racconti biografici. *Dossier Europa Emigrazione* 9: 3–5.

Silverstein, P.A. 2005. Immigrant Racialization and the New Savage Slot: Race, Migration, and Immigration in the New Europe. *Annual Review of Anthropology* 34: 363–384.

Smidt, W. 2010. The Tigrinnya-Speakers Across the Borders. In *Borders & Borderlands as Resources in the Horn of Africa*, ed. D. Feyissa and M.V. Hoehne, vol. 61. Suffolk/Rochester: James Currey/Boydell and Brewer.

———. 2014. The Term Habäša—An Ancient Ethnonym of the "Abyssinian" Highlanders and Its Interpretations and Connotations. In *Multidisciplinary Views on the Horn of Africa*, Series: Studien zum Horn von Afrika, ed. H. Elliese, vol. 1, 37–71. Koln: Kopp-Verlag.

Thomassen, B. 2014. *Liminality and the Modern: Living Through the In-Between.* Farnham: Ashgate.
Tibebu, T. 1996. Ethiopia: The "Anomaly" and "Paradox" of Africa. *Journal of Black Studies* 26 (4): 414–430.
Tronvoll, K. 2009. *War and the Politics of Identity in Ethiopia. The Making of Enemies & Allies in the Horn of Africa.* London: James Currey.
Turner, V. 1974. Liminal to Liminoid, in Play, Flow, and Ritual: An Essay in Comparative Symbology. *Rice Institute Pamphlet-Rice University Studies* 60 (3): 53–92.
Wacquant, L.J. 1997. Three Pernicious Premises in the Study of the American Ghetto. *International Journal of Urban and Regional Research* 21 (2): 341–353.
Wikan, U. 2013 [1992]. *Resonance: Beyond the Words.* Chicago: University of Chicago Press.

Index[1]

[1]Note: Page numbers followed by 'n' refer to notes.

G. Proglio et al. (eds.), *The Black Mediterranean*, Mediterranean Perspectives, https://doi.org/10.1007/978-3-030-51391-7

Ingram Content Group UK Ltd.
Milton Keynes UK
UKHW021425150623
423489UK00002B/156

9 783030 513900